STARTING STATISTICS

I Gwyn Evans

WITHDRAWN FROM STOCK

HODDER AND STOUGHTON
LONDON SYDNEY AUCKLAND TORONTO

Evans, I. Gwyn
Starting Statistics
1. Statistical mathematics—For schools
I. Title
519.5

ISBN 0 340 40733 6

First published 1988

Copyright © 1988 I. Gwyn Evans

All rights reserved. No part of this publication may be reproduced
or transmitted in any form or by any means, electronic, or
mechanical, including photocopy, recording, or any information
storage and retrieval system, without permission in writing from
the publisher or under licence from the Copyright Licensing
Agency Limited. Details of such licences (for reprographic
reproduction) may be obtained from the Copyright Licensing
Agency Limited, of 33–34 Alfred Place, London, WC1E 7DP.

Printed in Great Britain
for Hodder and Stoughton Educational
a division of Hodder and Stoughton Ltd, Mill Road
Dunton Green, Sevenoaks, Kent

Contents

Preface *iv*

Notation *iv*

1 Introducing data and variables *1*

Introduction; two examples of data collected by a class of pupils; variables; levels of measurement*.

2 Data on a qualitative variable *6*

Tabular presentation; diagrammatic presentation; non-frequency data; comparing two sets of data*. Projects. Review Exercises.

3 Small numerical data sets *17*

Diagrams; summary measures; working origins and scales*; transforming to given mean and standard deviation*; comparing examination marks for different subjects*. Projects. Review Exercises.

4 Discrete frequency distributions *33*

Frequency distributions; summary measures; graphical method for determining quartiles, deciles and percentiles; distributional shapes. Projects. Review Exercises.

5 Grouped frequency distributions *48*

Measurement accuracy; grouped data; modal class; cumulative frequency diagrams; percentiles using linear interpolation*; estimating the mean and standard deviation; groupings having unequal class intervals*. Projects. Review Exercises.

6* Comparing numerical data sets *67*

Introduction; diagrammatic comparisons; comparing summary measures; coefficient of variation. Review Exercises.

7 Populations, samples, and questionnaires *75*

Populations and samples; random samples; other random sampling procedures; sampling frame; some non-random sampling procedures; questionnaires; an example. Projects. Review Exercises.

8 Some other summary measures *86*

Weighted means; simple index numbers; weighted index numbers; geometric mean*; crude rates and standardized rates*. Review Exercises.

9 Paired variables *103*

Introduction; scatter diagram; line fitting; finding the slope of a fitted line; the line of 'best' fit*; correlation and Spearman's rank correlation coefficient; the product moment correlation coefficient*. Projects. Review Exercises.

10 Time series *126*

Time series data; seasonal variation; moving averages*. Review Exercises.

11 Probability *140*

Introduction; equally likely outcomes: elementary events; non-elementary events; tree diagrams; mutually exclusive events; probability tree diagrams; some further examples; relative frequency interpretation. Projects. Review Exercises.

12 Pictograms, misleading diagrams, ambiguous statements, and misinterpretations *155*

Enhanced diagrams; misleading diagrams; ambiguous statements; misinterpretations. A final word.

Answers to exercises *162*

Index *170*

* See final paragraph of Preface.

Preface

This text provides comprehensive coverage of all topics that may be included in any course on Descriptive Statistics and Elementary Probability. Some of the topics are relevant to pupils studying subjects such as Economics, Geography, Biology and Social Studies. The choice of topics has largely been governed by the latest GCSE syllabuses.

These syllabuses in Statistics have been developed so that pupils may meet and have a better understanding of basic statistical concepts and methodology. To achieve these aims each syllabus requires every pupil to carry out project work in order to gain some hands-on experience with the collection, presentation, analysis and interpretation of statistical data. The syllabuses are very much geared towards the role of Statistics in society with particular reference to its relevance to social, economic, and environmental factors.

Pupils and teachers will need to refer to their particular board's syllabus to determine which of the topics in the book are relevant.

Each chapter contains several worked examples to illustrate concepts and methodology. Exercises are interspersed throughout each chapter, and at the end of most chapters there are some suggested projects for pupils to do and review exercises graded into two levels; the answers to the exercises are given at the end of the book. It should be noted that some margin of variation should be allowed for answers that are obtained graphically.

Sections covering topics that are exclusive to the highest GCSE level (referred to by the various area boards as Level 2, Level C, Higher Level or Extended Level) are marked with an asterisk(*).

Notation

$=$	is equal to
$\approx$	is approximately equal to
$\neq$	is not equal to
$<$	is less than
$\leq$	is less than or equal to
$>$	is greater than
$\geq$	is greater than or equal to
$\sqrt{\ }$	the positive square root
Σ	the sum of
x	a variable or an arbitrary value of a variable
f	the frequency of a variable value
n	the number of observations in a data set
CF	cumulative frequency
CRF	cumulative relative frequency
$\bar{x}$	the (arithmetic) mean of a data set
m	the median of a data set
MAD($\bar{x}$)	the mean absolute deviation from the mean of a data set
MAD(m)	the mean absolute deviation from the median of a data set
VAR	the variance of a data set
SD	the standard deviation of a data set
LQ	the lower quartile of a data set
UQ	the upper quartile of a data set
SIQR	the semi-interquartile range of a data set
P_r	the rth percentile of a data set
CV	the coefficient of variation of a data set
WM	the weighted (arithmetic) mean
GM	the geometric mean of a data set
WGM	the weighted geometric mean
r_S	Spearman's rank correlation coefficient
r	the product moment correlation coefficient
P(A)	the probability of the event A occurring

1

INTRODUCING DATA AND VARIABLES

"I keep six honest serving-men
(They taught me all I knew);
Their names are What and Why and When
And How and Where and Who."

Rudyard Kipling, 'The Elephant's Child',
Just-So Stories (1902)

1.1 Introduction

The subject of Statistics is concerned with the collection and interpretation of data. It is relevant in many areas of human activity, including administration, advertising, scientific research, market research, and industry. Many government departments and local authorities keep records of data on human factors such as births, deaths, marriages, employment, housing, and so on. In industry, records are kept of data on production volumes, orders, sales, accidents, absenteeism, and the quality of the products. In sport, records are kept of data on the performances of individuals and teams. There are countless other areas in which data are collected. The data are often available in various publications. Data on human and other factors relating to the U.K. are published by Her Majesty's Stationery Office (H.M.S.O.). Some of the more useful H.M.S.O. publications include *Key Statistics*, *Abstract of Statistics* and *Social Trends*, which are published annually, and *The Digest of Statistics* and *Economic Trends*, which are published monthly. Data from these and other sources will be used throughout this book.

In the next section we give two examples of data collected by pupils in a class. The data will be used in later chapters to illustrate various statistical techniques.

1.2 Two Examples of Data Collected by a Class of Pupils

○ **EXAMPLE 1.1**

'Smarties' are sweets that are often sold in cylindrical tubes. To investigate possible variations in the tubes and their contents, each of 36 pupils in a class was given a small tube of Smarties and asked to provide the following information:

[1

(1) the colour of the cap (the lid of the tube),
(2) the letter on the inside face of the cap,
(3) the number of sweets in the tube,
(4) the total weight, to the nearest 0·5 gram, of the sweets in the tube.

(Chemical weighing scales were borrowed to weigh the sweets. Under current weights and measures regulations the weight of the contents of a tube, being less than 50 grams, does not have to be stated on the tube.)

The data obtained are displayed in Table 1.1.

Table 1.1 *Information on 36 Tubes of Smarties*

Tube	Cap colour	Cap letter	Number of sweets	Weight of sweets (g)
1	Blue	g	37	35.0
2	Orange	c	38	35.5
3	Green	m	40	37.0
4	Yellow	d	37	34.0
5	Green	v	39	36.0
6	Orange	h	38	35.0
7	Orange	p	37	34.5
8	Orange	a	39	36.5
9	Green	t	38	35.5
10	Blue	d	36	33.5
11	Orange	i	37	35.5
12	Blue	x	38	35.5
13	Orange	g	36	34.0
14	Yellow	z	39	36.5
15	Green	y	38	35.5
16	Yellow	b	37	34.0
17	Orange	p	38	35.0
18	Green	k	39	36.0
19	Green	g	37	35.0
20	Orange	t	38	35.0
21	Blue	c	37	34.5
22	Orange	q	38	36.0
23	Green	e	38	36.0
24	Orange	w	36	34.0
25	Orange	b	39	36.5
26	Blue	p	38	35.5
27	Yellow	h	38	36.0
28	Green	i	38	36.0
29	Orange	r	37	34.5
30	Blue	d	37	35.0
31	Blue	t	37	35.0
32	Orange	m	39	36.5
33	Orange	w	38	35.5
34	Yellow	p	40	37.0
35	Orange	k	37	34.5
36	Green	g	37	35.0

○ **EXAMPLE 1.2**

The same 36 pupils were also asked to give the following information about themselves:

(1) their sex,
(2) their method of travel to school,
(3) their month of birth,
(4) their shoe size,
(5) the number of boys in their family,
(6) the number of girls in their family,
(7) their height, to the nearest centimetre,
(8) their weight, to the nearest 0·1 kilogram.

The data obtained are displayed in Table 1.2.

1.3 Variables

When given a table such as Table 1.1 or 1.2, the first thing to do is to look at what has been tabulated. The heading of Table 1.1 tells us that the data relate to 36 tubes of Smarties. The individual column headings show that four different characteristics associated with a tube have been examined. These characteristics are (1) the colour of the cap, (2) the letter on the inside of the cap, (3) the number of sweets in the tube, (4) the weight of the sweets in the tube. Each such characteristic is an example of a **variable** associated with a tube of Smarties. It is so called because what is observed varies from one tube to another.

In Table 1.2 the variables observed are (1) sex, (2) method of travel to school, (3) month of birth, (4) shoe size, (5) number of boys in the family, (6) number of girls in the family, (7) height, and (8) weight.

A variable whose values are numerical is said to be a **quantitative variable**. The quantitative variables in Example 1.1 are:

(3) the number of sweets in a tube,
(4) the weight of the sweets in a tube.

The quantitative variables in Example 1.2 are:

(4) the size of shoe worn by a pupil,
(5) the number of boys in a pupil's family,
(6) the number of girls in a pupil's family,
(7) the height of a pupil,
(8) the weight of a pupil.

● *EXERCISE 1.1*

1 State which of the following variables are quantitative.
(a) A person's religion.
(b) The number of male mice in a litter of mice. ✓
(c) The spelling ability of a child, measured as

Table 1.2 *Information on 36 Pupils in a Class*

Pupil	Sex	Method of travel	Month of birth	Shoe size	Family size: boys	Family size: girls	Height (cm)	Weight (kg)
1	M	Bus	January	5	1	0	159	54.8
2	M	Car	April	$5\frac{1}{2}$	1	2	167	59.5
3	F	Bus	June	5	0	1	152	49.3
4	F	Walk	February	6	0	2	169	62.6
5	M	Bus	May	$6\frac{1}{2}$	2	0	170	62.2
6	M	Walk	August	$6\frac{1}{2}$	2	2	170	59.9
7	M	Bus	March	$7\frac{1}{2}$	4	1	174	66.7
8	M	Other	November	8	3	0	173	64.7
9	F	Bus	June	$4\frac{1}{2}$	0	2	147	51.0
10	F	Bus	April	$6\frac{1}{2}$	3	1	169	63.6
11	F	Other	August	5	2	2	150	53.7
12	M	Bus	May	$5\frac{1}{2}$	2	1	162	56.3
13	M	Walk	June	5	1	0	162	58.1
14	F	Walk	July	4	1	1	147	49.5
15	F	Bus	April	$5\frac{1}{2}$	2	2	163	59.5
16	F	Car	May	4	1	2	142	44.0
17	M	Walk	December	$7\frac{1}{2}$	2	0	171	62.2
18	F	Bus	February	$4\frac{1}{2}$	2	4	145	46.4
19	M	Other	June	6	3	0	166	58.8
20	F	Walk	August	5	1	1	148	46.1
21	M	Bus	April	7	3	1	172	65.4
22	M	Bus	July	$6\frac{1}{2}$	2	0	168	59.0
23	F	Car	October	5	2	1	157	56.7
24	M	Walk	March	$5\frac{1}{2}$	1	2	165	59.5
25	M	Walk	May	6	1	0	166	56.8
26	M	Car	September	5	1	3	163	57.9
27	M	Walk	July	7	2	0	170	64.9
28	F	Bus	October	$5\frac{1}{2}$	4	1	155	55.6
29	F	Bus	January	$5\frac{1}{2}$	0	2	160	58.0
30	M	Car	July	$7\frac{1}{2}$	2	1	172	61.3
31	M	Other	June	$7\frac{1}{2}$	2	0	173	68.6
32	F	Bus	November	$6\frac{1}{2}$	1	2	170	64.1
33	F	Car	May	5	0	4	145	48.5
34	F	Bus	October	$4\frac{1}{2}$	1	2	154	53.0
35	M	Bus	September	8	1	2	175	68.4
36	M	Walk	May	7	2	0	171	62.9

how many of 50 words the child can spell correctly.
(d) The extension in the length of a heated metal rod.
(e) An athlete's position at the end of a race.
(f) The number of roses in a flower bed.
(g) A boy's rating of how pretty a girl is.
(h) The amount a family spend on food in a week.
(i) The number of heads obtained in ten tosses of a coin.
(j) The time taken by an athlete to complete a race over a given distance.
(k) The colour of a postage stamp.
(l) The rateable value of a house.
(m) The number of bedrooms in a house.
(n) The size of a girl's dress, or of a boy's shirt.

A quantitative variable is said to be **discrete** if it is restricted to certain values. For example, in Table 1.1 the variable 'number of sweets in a tube' must have a value which is a whole number. Since the whole numbers can be written down individually this is an example of a discrete variable. In fact this is the only discrete variable in Table 1.1. In Table 1.2 the discrete variables (each of whose possible values can be listed individually) are: 'shoe size', 'the number of boys in the family', and 'the number of girls in the family'.

Variables such as 'weight of sweets' in Table 1.1. and 'height' and 'weight' in Table 1.2 are different in that their possible values are less restricted and we cannot list all the possible values one by one. To appreciate this, try making a list of all possible heights of 15-year old girls, say. Such variables are said to be **continuous**. In most cases the value of a continuous variable has to be determined using some sort of measuring device (for example, a ruler or tape measure when finding a length or a height, weighing scales when finding the weight of an object).

2 State which of the quantitative variables in Question 1 are (i) discrete, (ii) continuous.

A variable whose values are not numerical is said to be a **qualitative variable**. Such a variable will have values which are verbal descriptions rather than numbers; the various descriptions are often referred to as **categories**. For example, the variable 'cap colour' in Table 1.1 is qualitative, having the values 'blue', 'orange', 'green', and 'yellow'. The variable 'cap letter' in Table 1.1 is also qualitative. In Table 1.2 the qualitative variables are 'sex', 'method of travel to school', and 'month of birth'.

3 State which of the variables in Question 1 are qualitative.

4 Table 1.3 gives information on 10 new cars available for sale at a garage. Write down the variables in this table that are (a) qualitative, (b) discrete, (c) continuous.

Table 1.3

Car	Colour	Type	Number of doors	Engine capacity (cc)	Maximum speed (kph)
1	Red	Saloon	4	1296	140
2	Red	Hatchback	3	1598	172
3	White	Hatchback	5	1586	168
4	White	Saloon	2	1595	170
5	Blue	Estate	5	1986	188
6	Blue	Saloon	4	1298	145
7	Blue	Hatchback	3	1595	172
8	Blue	Saloon	2	1312	150
9	Yellow	Saloon	4	1625	172
10	Black	Estate	3	1597	167

Measuring feelings

The following style of question is quite common in attitude or opinion questionnaires:

For each of the following items put a ring round the degree of agreement or disagreement which you feel for the statement.

	Strongly disagree	Disagree	Indifferent	Agree	Strongly agree
1 Chewing gum helps me to concentrate	−2	−1	0	+1	+2
2 'Bubble-seven' soap is better than its rivals	−2	−1	0	+1	+2
3 It is time for a change of government	−2	−1	0	+1	+2

If John and Mary have to reply to this type of questionnaire they might give the following responses:

	John	Mary
Item 1	+2	+1
2	−1	+1
3	+1	+1

For item 1, what statistical assumptions would have to be true for us to say that "John is twice as much in agreement as Mary"? (Units and absolute zero.) Would it be fair to averge all John's responses to measure his general level of agreement? Even if all the items were on the same topic, which these are not, what statistical assumptions across the items would need to be made? (Units and absolute zero.) If we cannot get an overall measure of John's agreement to compare with that of Mary's, how could we deal with the data?

1.4 Levels of Measurement*

Variables may also be classified according to the level of sophistication of the 'measured' values. In particular we consider whether two values of a variable can be sensibly compared by:

(1) subtracting one from the other,
(2) dividing one by the other.

First, consider a quantitative variable such as weight. Given that one object has a weight of 15 grams and another has a weight of 5 grams, then it is certainly correct to say that:

(1) the first object weighs $(15-5)$ g $= 10$ g more than the second object,
(2) the first object weighs $15/5 = 3$ times as much as the second object.

The same is true for variables such as 'height' and 'number of sweets in a tube of Smarties', and, indeed, for any variable whose values are obtained relative to a true zero point. Such variables are said to be measured on a **ratio scale**.

There are some quantitative variables that are not measured on a ratio scale. An example of a continuous variable for which this is so is the hotness or temperature of an object. Temperatures can be measured on the Celsius scale in degrees Celsius (°C). The zero point (0°C) of the Celsius scale has been arbitrarily chosen to be the temperature at which water freezes. However, objects can be cooled below this zero point. For a true zero point when finding the temperature of an object we have to use the absolute (or Kelvin) temperature scale. The zero point on this scale, written 0 K, is equivalent to about -273°C on the Celsius scale.

Consider two objects, one having a temperature of 20°C and the other a temperature of 60°C. Does this mean that (i) the second object is 40° hotter than the first object, (ii) the second object is three times as hot as the first object? On the absolute temperature scale (measured in Kelvins), the first object has a temperature of $20+273 = 293$ K, while the second object's temperature is $60+273 = 333$ K. Thus the true difference in the hotness of the objects is $333-293 = 40$, exactly as on the Celsius scale. The true ratio of the two temperatures is $333/293 \approx 1.14$, which is very different from the ratio 3 obtained from the temperature as measured on the Celsius scale. It follows that we can meaningfully subtract temperatures on the Celsius scale, but dividing temperatures on this scale is not meaningful because it does not give the true ratio of two temperatures.

A variable whose values may be subtracted meaningfully but not divided meaningfully is said to be measured on an **interval scale**. As we have seen, temperature measured on the Celsius scale is a variable measured on an interval scale; the Fahrenheit scale is also an interval scale.

Shoe size is an example of a discrete variable which is measured on an interval scale. Shoe sizes usually go up in halves, an increase in size by $\frac{1}{2}$ corresponding to an increase of about 4 millimetres in length. Subtracting one size from another gives an indication of the difference in the lengths of the shoes. For example, the difference in the lengths of shoes of sizes 6 and 8 is the same as that for shoes of sizes 4 and 6. However, a shoe of size 8 is not twice the length of a shoe of size 4. Here again we have a variable whose values are measured with respect to an arbitrarily chosen zero, namely the length of a shoe which is of size 0 on the size scale.

● **EXERCISE 1.2**

1 State which of the variables in Question 1, Exercise 1.1 are measured on (i) a ratio scale, (ii) an interval scale.

Qualitative variables also have two levels of measurement. A qualitative variable whose possible values (categories) can be ranked according to magnitude is said to be measured on a **rank** or **ordinal scale**. We have not yet met an example of such a variable. The following are examples of qualitative variables measured on a rank scale:

(1) the social status of a person, possible categories being 'upper class', 'middle class' and 'lower class',
(2) coffee taste, possible categories being 'nice', 'satisfactory', and 'unpleasant',
(3) results of a horse race, possible categories being 'first', 'second', 'third', and 'also ran',
(4) examination grades, possible categories being A, B, C, D, E, and U.

In each of these examples there is a natural ordering of the categories. In the first example an 'upper class' person is deemed to be wealthier than a 'middle class' person, who, in turn, is deemed to be wealthier than a 'lower class' person. But in none of the examples is it possible to quantify the difference between two categories.

A qualitative variable whose possible values (categories) cannot be ranked is said to be measured on a **nominal scale**. Examples from Tables 1.1 and 1.2 are as follows:

(1) cap colour,
(2) cap letter,
(3) sex,
(4) method of travel to school,
(5) month of birth.

2 State which of the variables in Question 1, Exercise 1.1 are measured on (i) a rank scale, (ii) a nominal scale.

2 DATA ON A QUALITATIVE VARIABLE

"This is my data —
What shall I do with it?!!"
Anon.

In this chapter we consider methods of presentation and analysis of data on a qualitative variable.

2.1 Tabular Presentation

Given observations of a qualitative variable the first step is to put the data in tabular form as illustrated in the following example.

○ EXAMPLE 2.1

Consider the observations on cap colour in Table 1.1. p. 2. As presented in the table, it is difficult to appreciate the variation in the colours of the caps of the tubes. For a clearer picture we first count the number of caps of each colour. This is conveniently done by means of a **tally chart**.

Table 2.1 *Tally Chart for Cap Colour in Table 1.1*

Cap colour	Tally	Frequency
Blue	JHT II	7
Orange	JHT JHT JHT	15
Green	JHT IIII	9
Yellow	JHT	5
	Total	36

The tally chart for cap colour is shown in Table 2.1. To obtain this table, read down the appropriate column of Table 1.1. The colour of the first cap is blue, so write down the word 'blue' and alongside it place a vertical stroke, as shown in the first and second columns of Table 2.1. The colour of the cap of the second tube in Table 1.1 is orange, so write 'orange' beneath 'blue' in the first column of Table 2.1 and place a vertical stroke alongside it in the second column. Continue in this way for the cap colours of all 36 tubes, placing a vertical stroke alongside each colour as it arises. Every fifth stroke in any row should be drawn across the preceding four strokes. This makes it easier to count the number of strokes in each row. The total number of times that a particular variable value occurs is called the **frequency** of that value. The frequencies of the various cap colours in our example are given in the last column of Table 2.1. The total of the frequencies has been shown as a check that all 36 tube have been covered.

Having compiled the tally chart, a more compact presentation of the cap colours is that shown in Table 2.2. This table is referred to as the **frequency distribution** of the cap colours. It shows that orange is the colour that occurred most frequently. The most frequently occurring value of a qualitative variable is called its **modal value**. In our example the modal colour is orange.

Table 2.2 *Frequency Distribution of Cap Colour from Table 1.1*

Cap colour	Blue	Orange	Green	Yellow	Total
No. of caps (frequency)	7	15	9	5	36

If the cap colours were produced in equal numbers we would expect about $\frac{1}{4} \times 36 = 9$ caps of each of the four colours. Since the number of orange caps (15) is three times the number of yellow caps (5) it seems unlikely that the colours are produced in equal numbers, unless, for some reason, our sample of 36 tubes is not representative of all the tubes produced.

○ **EXAMPLE 2.2**

Now consider the letters on the caps. With only 36 tubes there is little point in looking at all 26 letters of the alphabet. Instead, let us classify the letters as being vowels (a,e,i,o,u) or consonants (all other letters). From the entries in Table 1.1, the frequency distribution (which you may like to verify) of our letter classification is shown in Table 2.3.

Table 2.3 *Frequency Distribution of Letter Type from Table 1.1*

Letter type	Vowel	Consonant	Total
No. of caps (frequency)	4	32	36

We know that only 5 of the 26 letters of the alphabet are vowels. If all the letters were produced in equal numbers then of our 36 tubes we would expect about $\frac{5}{26} \times 36 = 7$, to the nearest whole number, to be vowels. We have 4 vowels, which does not differ very much from 7. Equivalently our observed 32 consonants does not differ very much from the expected number 29. The discrepancy is not really large enough to rule out the possibility that all 26 letters are produced in equal numbers.

But wait! One possible purpose of the letters on the caps is for children to make up words. Studies of English prose have shown that about 38% of the letters used are vowels. So to be useful for word making about 38% of the letters on the caps should be vowels. For 36 tubes we would then expect the number of vowels to be about 38% of 36, which is almost 14. We only have half this number. So if our sample is fairly representative of all the letters produced we can conclude that it is most unlikely that the aim is to make word-making easier.

● **EXERCISE 2.1**

1 Referring to Table 1.2, p. 3, draw up a tally chart for the methods of travelling to school used by the 36 pupils. Present the results as a frequency distribution table. State the modal method of travel. (Keep your table for use in Exercise 2.2.)

2 Referring to the months of birth of 36 pupils given in Table 1.2, draw up a tally chart for the seasons of birth, taking the seasons to be Spring (March–May), Summer (June–August), Autumn (September–November), and Winter (December–February). Draw up a table of the frequency distribution of the seasons of birth. Discuss the possibility that the numbers of children born in the four seasons are roughly equal.

3 Refer to the opening sentence of this chapter, which starts off: 'In this chapter we consider . . .'. Preferably by means of a tally chart, record the numbers of vowels and consonants in the sentence. Compare the percentage of letters that are vowels with the value 38% for English prose.

4 Repeat Question 3, but this time use two or three lines (aiming for at least 100 letters) from any novel that you have.

5 With reference to Question 4, Exercise 1.1, p. 4, draw up a frequency distribution table for (a) car colour, (b) car type. In each case state the modal value of the variable. (Keep your tables for use in Exercise 2.2.)

2.2 Diagrammatic Presentations

Representing data diagrammatically provides a simple way of conveying some features of the data. In this section we shall illustrate diagrams that are appropriate for data on a qualitative variable, using the data on cap colour as presented in Table 2.2.

☐☐ BAR DIAGRAMS

A simple form of diagram for displaying the frequency distribution in Table 2.2 is a **bar diagram**, sometimes referred to as a **block diagram** or as a **bar chart**. In such a diagram a rectangle is drawn to represent the frequency of a variable value (category).

A bar diagram for Table 2.2 is shown in Figure 2.1a. The variable values (cap colours) have been ordered along the variable axis as they appear in Table 2.2. An alternative is shown in Figure 2.1b. Here the variable values have been ordered along the variable axis, starting with the one having the highest frequency and ending with the one having the lowest frequency. Another alternative is to order the variable values alphabetically along the variable axis.

Although the choice of how to order the variable values is not crucially important, many people think that ordering by frequency, as in Figure 2.1b, gives a clearer picture of the distribution. (You judge for yourself.) Whichever ordering is chosen, the following points should be observed when constructing a bar diagram:

(1) One axis (usually the horizontal one) is used for the variable, and the other axis (the vertical one) is used for the frequency; the frequency axis must be scaled appropriately and clearly.
(2) All the rectangles must be the same width, and their heights must be proportional to the frequencies they represent. Then the areas of the rectangles will also be proportional to the frequencies. This is important because people judge the size of a rectangle, and hence the frequency it represents, by its area, not just its height.
(3) For neatness of presentation the gaps between the rectangles should be equal.

☐☐ COMPONENT BAR DIAGRAMS

An alternative form of diagrammatic presentation of Table 2.2 is shown in Figure 2.2, which is referred to as a **component bar diagram**, or sometimes as a **composite bar diagram**. In this diagram a rectangle has been drawn with area (and equivalently, length) proportional to the total frequency (36) and has been subdivided into rectangles with areas (and lengths) proportional to the frequencies. Here we have chosen to order the cap colours according to the magnitudes of their frequencies. The frequencies are shown in brackets to make the diagram easier to read.

Figure 2.2 Component Bar Diagram for Table 2.2

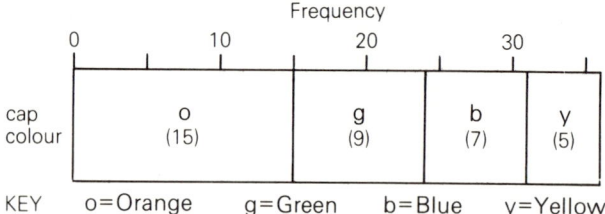

Figure 2.1 Bar Diagrams for Table 2.1. (a) Colours Ordered as in Table, (b) Colours Ordered by Frequency

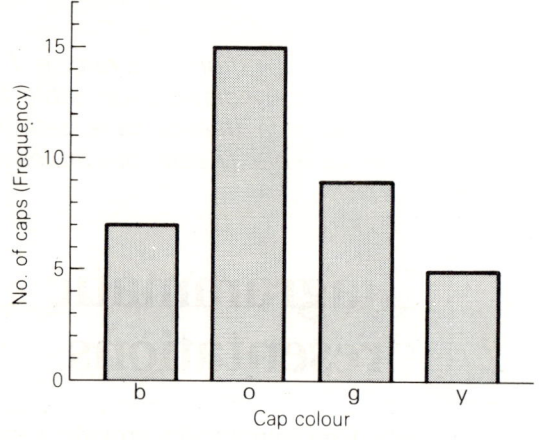
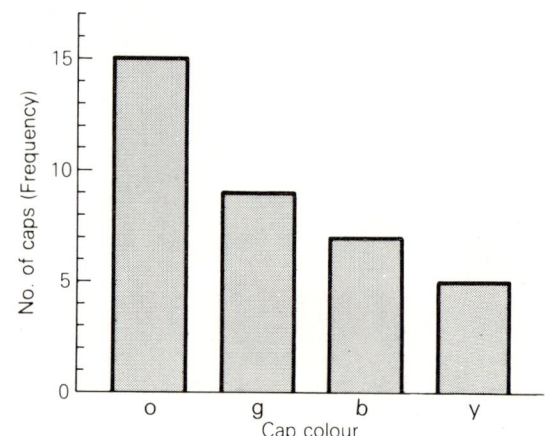

KEY: b=Blue o=Orange g=Green y=Yellow

□□ PIE CHARTS

A popular variant of the component bar diagram is one in which a circle is used instead of a rectangle. The circle is drawn having area proportional to the total frequency. It is then subdivided into sectors having areas proportional to the frequencies of the variable values. Such a diagram is known as a **pie chart**, or sometimes as a **circular diagram**. A pie chart for the frequency distribution of the cap colours in Table 2.2 is shown in Figure 2.3. The various sectors are obtained by subdividing the total angle of 360° at the centre of the circle into angles that are proportional to the frequencies of the variable values (cap colours). In our example the total frequency is 36, so that each unit of frequency (i.e. each of the 36 observations) is assigned an angle of 360°/36 = 10°. Thus the sector angles for the various colours are:

Blue: $7 \times 10° = 70°$
Orange: $15 \times 10° = 150°$
Green: $9 \times 10° = 90°$
Yellow: $5 \times 10° = 50°$

Figure 2.3 *Pie Chart for Table 2.2*

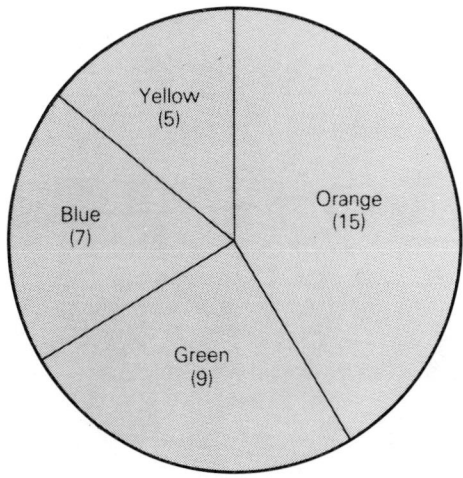

As in our previous diagrams we have elected to order the colours according to their frequencies, in this case in a clockwise direction starting from the position of '12 o'clock' in the belief that this particular ordering gives a better visual representation. We have also included, in brackets, the frequency of each colour, so that all the information in Table 2.2 is contained in the diagram.

● *EXERCISE 2.2*

1 In Question 1, Exercise 2.1, you were asked to compile the frequency distribution table for the methods of travelling to school used by 36 pupils. Represent this distribution in the form of (a) a bar diagram, (b) a component bar diagram, and (c) a pie chart.

2 Refer to Table 1.2, p. 3, and draw a pie chart to represent the methods of travelling to school used by the 20 boys in the class.

3 In Question 5, Exercise 2.1, you were asked to compile the frequency distribution tables of the colour and type of 10 cars. Represent the distribution of car colour in the form of a bar diagram and that of car type in the form of a pie chart.

In the pie chart example (Figure 2.3) calculation of the sector angles was very easy because the total frequency of 36 divides exactly into 360°. The same remark applies to the pie charts you were asked to draw in Exercise 2.2. When the total frequency does not divide exactly into 360° calculation of the sector angles is a little more complicated, as illustrated in the following example.

○ *EXAMPLE 2.3*

Table 2.4, extracted from Table 7.17 of the 1985 Edition of *Social Trends*, shows the numbers of males who died from various types of accidents in homes and residential accommodation in 1983.

Table 2.4 *Male Deaths from Accidents in Home or Residential Accommodation in 1983*

Accident type	No. of deaths
Poisoning	322
Fall	1039
Fire	351
Suffocation	255
Other	248
Total	2215

To construct a pie chart for these data we first observe that the total frequency is 2215, so that each unit of frequency has to be assigned an angle of $(360/2215)°$. Then the sector angles for the accident types calculated to the nearest degree are obtained as follows:

Poisoning: $322 \times \frac{360}{2215} = 52°$

Fall: $1039 \times \frac{360}{2215} = 169°$

Fire: $351 \times \frac{360}{2215} = 57°$

Suffocation: $255 \times \frac{360}{2215} = 41°$

Other: $248 \times \frac{360}{2215} = 40°$

Total   359°

The fact that the sector angles add to only 359° instead of 360° is due to the slight errors introduced in evaluating each angle correct to the nearest degree. Calculating angles to the nearest degree is sufficiently accurate for diagrammatic purposes. The corresponding pie chart is given in Figure 2.4, where we have again ordered the categories clockwise according to their frequencies, starting from '12 o'clock'.

Figure 2.4   *Pie Chart for the Data in Table 2.4*

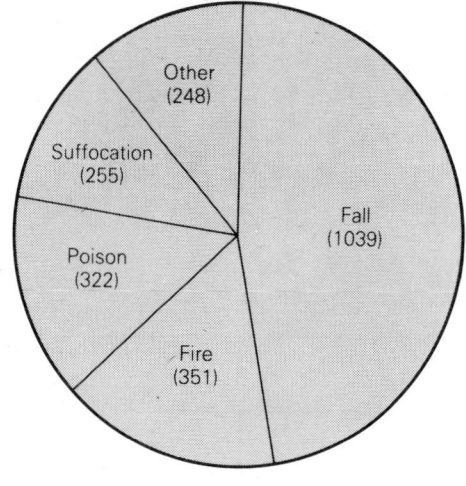

4   *Referring to Table 1.2, p. 3, construct a pie chart to illustrate the methods of travelling to school used by the 16 girls in the class.*

Table 2.5

Magazine	Readership (m)
Woman's Own	5.2
Woman	4.3
Woman's Weekly	3.7
Family Circle	3.4

5   *Table 2.5, extracted from Table 10.8 of the 1986 Edition of Social Trends, shows the numbers of readers of selected women's magazines in 1984 in millions (m). Draw a pie chart for the data.*

2.3 Non-Frequency Data

In the example and exercises given so far counts have been made of the number of individual observations (frequencies) in each category of a qualitative variable. The diagrammatic representations used are equally valid when some quantitative measurement other than a count is associated with the different categories of a qualitative variable. As an example consider the data displayed in Table 2.6 which are the expenditures (in millions of pounds) by the British Council in 1984/85 in the provision of overseas aid in various areas of education and training.

Table 2.6   *British Council Expenditures on Overseas Aid in 1984/85*

Area of expenditure	Expenditure (£m)
English Language	40.2
Science	22.1
Education	18.4
Arts	13.1
Media	3.9

Figure 2.5   *Bar Diagram for Table 2.6*

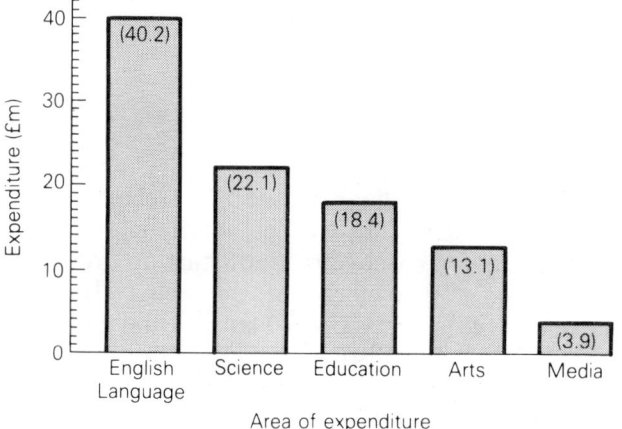

Figure 2.5 is a bar diagram illustrating these data.
A component bar diagram or a pie chart, may also be used to display data such as that in Table 2.6. For the pie chart we first evaluate the total expenditure in millions of pounds, which is:

$$40{\cdot}2 + 22{\cdot}1 + 18{\cdot}4 + 13{\cdot}1 + 3{\cdot}9 = £97{\cdot}7\text{m},$$

so that each million pounds is assigned an angle of $(360/97{\cdot}7)°$. The sector angles for the various areas of expenditure are then calculated as follows.:

English Language: $40.2 \times \dfrac{360}{97.7} = 148°$

Science: $22.1 \times \dfrac{360}{97.7} = 81°$

Education: $18.4 \times \dfrac{360}{97.7} = 68°$

Arts: $13.1 \times \dfrac{360}{97.7} = 48°$

Media: $3.9 \times \dfrac{360}{97.7} = 14°$

Each angle has been given to the nearest degree. Here, as in the previous pie chart example, the angles add to 359°, the discrepancy again resulting from rounding the angles to the nearest degree. The corresponding pie chart is displayed in Figure 2.6.

Figure 2.6 *Pie Chart for British Council Expenditures on Overseas Aid in 1984/85*

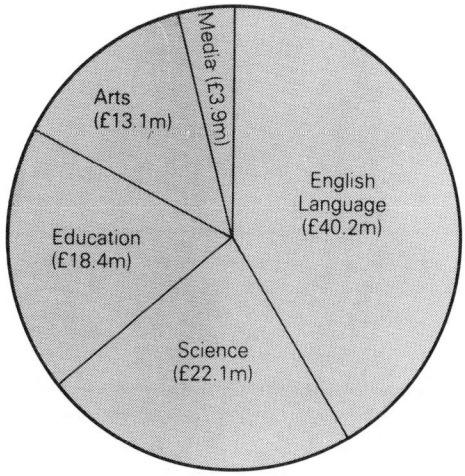

● **EXERCISE 2.3**

For each of the following sets of data draw (a) a bar diagram, (b) a pie chart.

1 *At the beginning of a certain year the managers of Britain's Investment Trusts recommended the breakdown shown in Table 2.7 for an individual's investment.*

Table 2.7

Area of investment	Percentage
U.K.	46
North America	32
Japan	15
Other	7

2 *The areas of the four countries comprising the United Kingdom are: England 130 359 km²; Scotland 78 764 km²; Wales 20 761 km²; and Northern Ireland 14 118 km².*

3 *Table 2.8 shows the domestic consumptions in thousands of tonnes of solid fuels in a particular district in one year.*

Table 2.8

Type of solid fuel	House coal	Coke	Other
Consumption ('000 tonnes)	1660	256	84

2.4 Comparing Two Sets of Data on a Qualitative Variable*

Now we shall consider diagrams suitable for exhibiting similarities and differences in two sets of data on a qualitative variable. We shall illustrate the various possibilities using the data on the method of travelling to school used by 36 pupils, consisting of 20 boys and 16 girls, given in Table 1.2, p. 3. The separate distributions for boys and girls are shown in Table 2.9. (You may have derived these distributions in Questions 2 and 4 of Exercise 2.2).

☐☐ DUAL BAR DIAGRAMS

One way of representing the data in Table 2.9 is by means of a **dual bar diagram** as shown in Figure 2.7. In this diagram all the rectangles have been drawn the same width, and the pairs of rectangles for each category of travel are equally spaced along the variable axis.

The fact that the rectangles have equal widths can be misleading when making comparisons. For

[11

Table 2.9 *Method of Travelling to School Used by 20 Boys and 16 Girls*

Method of travel	Bus	Walk	Car	Other	Total
No. of boys	7	7	3	3	20
No. of girls	9	3	3	1	16

Figure 2.7 *Dual Bar Diagram for Table 2.9*

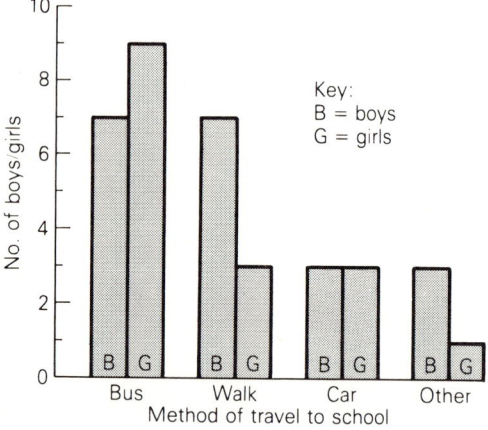

instance, the areas of the rectangles in Figure 2.7 suggest that more than twice as many boys as girls walk to school, and that three times as many boys as girls travel to school other than by bus, car or walking. Such comparisons take no account of the fact that there are more boys than girls in our sample.

To enable comparisons to be made between the breakdowns for boys and girls we need to work with **relative frequencies** rather than frequencies. The relative frequency of any variable value is its frequency divided by the total frequency. It is the proportion of all the observations that have that particular value. For example, since 7 of the 20 boys travel by bus, the relative frequency of boys travelling by bus is 7/20 = 0·35 or 35%; and since 9 of the 16 girls travel by bus, the corresponding relative frequency for girls is 9/16 = 0·5625 or 56%, to the nearest whole per cent. The other relative frequencies are calculated similarly, their values as percentages being shown in Table 2.10.

It is evident from Table 2.10 that a higher proportion of girls than boys travel to school by bus,

Table 2.10 *Relative Frequencies (as Percentages) for Table 2.9*

Method of travel	Bus	Walk	Car	Other
% of boys	35	35	15	15
% of girls	56	19	19	6

and that higher proportions of boys than girls walk to school, and travel by means other than bus, car or walking. A convenient diagram for showing these differences is the **dual percentage bar diagram**, which is given in Figure 2.8. This differs from Figure 2.7 only in having the vertical axis scaled in relative frequencies (expressed as percentages).

Figure 2.8 *Dual Percentage Bar Diagram for Table 2.10*

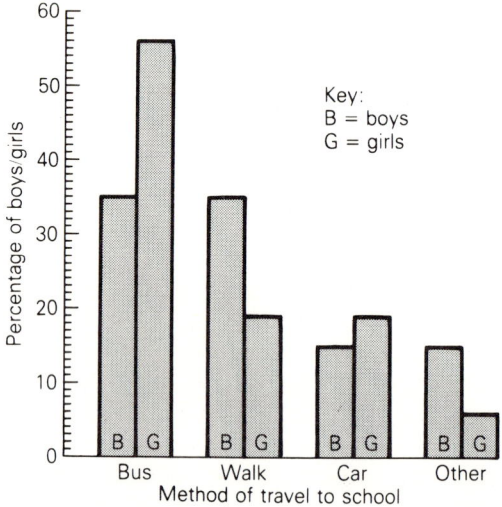

The dual percentage bar diagram serves well for a visual comparison of the various methods of travelling to school used by boys and girls. However, it does not give any indication at all about the numbers of boys and girls involved. For a diagram which will indicate the difference between the sizes of two data sets *and* the breakdowns of the variable values in the two data sets we may use either two component bar diagrams or two pie charts side by side. One diagram represents each data set, and both are constructed in such a way that the areas of the diagrams are proportional to the sizes of the two data sets. Let us illustrate the use of pie charts in this way for the data of Table 2.9. We need to draw a circle to represent the boys' data and another circle to represent the girls' data with *areas* proportional to the total frequencies of 20 for boys and 16 for girls, respectively.

Since the area of a circle is π times the square of its radius, the radii r_B and r_G of the boys' and girls' circles must be chosen so that:

$$\frac{\pi r_B^2}{\pi r_G^2} = \frac{20}{16},$$

or, cancelling out the π and taking square roots, so that:

$$\frac{r_B}{r_G} = \sqrt{\frac{20}{16}} = 1 \cdot 12,$$

to two decimal places. In Figure 2.9 we have drawn a circle of radius $r_G = 2 \cdot 5$ cm to exhibit the girls' data, and consequently the radius of the circle for the boys' data has to be $r_B = 1 \cdot 12 \times 2 \cdot 5 = 2 \cdot 8$ cm. The sector angles are calculated in the usual way as follows:

	Girls	Boys
Bus:	$9 \times \frac{360}{16} = 202 \cdot 5°$	$7 \times \frac{360}{20} = 126°$
Walk:	$3 \times \frac{360}{16} = 67 \cdot 5°$	$7 \times \frac{360}{20} = 126°$
Car:	$3 \times \frac{360}{16} = 67 \cdot 5°$	$3 \times \frac{360}{20} = 54°$
Other:	$1 \times \frac{360}{16} = 22 \cdot 5°$	$3 \times \frac{360}{20} = 54°$

Figure 2.9 provides direct visual information on the relative numbers of boys and girls in the sample, and by including the category percentages (shown in brackets and obtained from Table 2.7, those for girls being given to the nearest whole per cent), it also provides a visual comparison of the breakdowns by category for boys and girls.

Figure 2.9 *Comparative Pie Charts for the Data in Table 2.9*

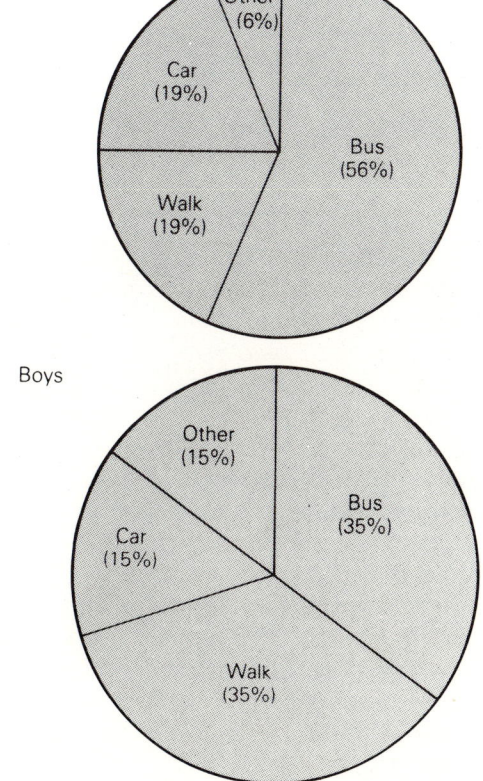

● **EXERCISE 2.4**

1 *90 boys and 40 girls at a certain school were asked to rate the sporting facilities at the school as Good, Satisfactory, or Poor. A total of 38 pupils rated the facilities as Good, and 69 rated them as Satisfactory; 15 of the boys rated them as Poor and 12 of the girls rated them as Good. Display this information in the form of a table. Draw two pie charts, one for the boys' ratings and one for the girls' ratings, suitable for comparing the two sets of responses.*

2 *Table 2.11 shows a motoring organisation's estimates of the running costs for two car models, in pounds, based on the first 10 000 miles of motoring in a certain year.*
Draw two pie charts, one for each model, appropriate for making comparisons of the two models.

Table 2.11

	Model A (£)	Model B (£)
Tax and insurance	180	220
Depreciation	780	800
Petrol and oil	550	650
Servicing	90	150
Total	1600	1820

3 *Table 2.12 shows the distribution according to marital status of males and females over 15 years old.*
Draw a dual bar diagram to illustrate these data.

Table 2.12

Marital status	Males (percentage of total)	Females (percentage of total)
Single	25.7	21.4
Married	69.7	64.0
Widowed	3.9	13.5
Divorced	0.7	1.1

What price experience?

Statistics on sex discrimination can be difficult to unravel. If we consider the following hypothetical example of college recruitment, we can see one of the problems.

	No. applied	No. accepted	% accepted
Arts Faculty			
Women	250	100	40%
Men	125	50	40%
Science Faculty			
Women	100	70	70%
Men	225	158	70%

Neither the Arts not the Science Faculty show signs of discrimination. However, if we had been given the total recruitment figures for the college, we might well have drawn a different conclusion.

	No. applied	No. accepted	% accepted
Women	350	170	49%
Men	350	208	59%

Data on sex discrimination in salary prospects are also often misinterpreted. It is always difficult to ensure comparisons of like with like. The comparison of average salaries for men and women in a particular job requires the identification of equivalent work in terms of duties and status. In recent years, there had been a rapid increase in the number of women entering the job market. Many of these are women who have taken time out to bring up families. In comparison with men of the same age but unbroken careers, the women may therefore have less (relevant) experience. For that reason they may command lower salaries, even though they are doing much the same work as the men.

*Projects

1 Visit a newsagent and request the names (titles) of the national daily newspapers sold there, and the number of copies of each sold the previous day (or any other day or week for which a record has been kept). Display the data in a table and in a diagram.

The national daily newspapers which have a small page size of approximately 38 cm by 30 cm are referred to as tabloids. Discuss the possibility that all the tabloids are equally popular among the newsagent's customers.

Taking account of the prices of the various newspapers do your data suggest that the number of copies sold depends on the price?

2 When a District Council sends out a demand to a house occupier for the payment of rates it usually includes a leaflet giving a breakdown of how much of each £1 paid in rates is earmarked for the provision of various services. Ask your parents if they have such a leaflet; if not, then visit the District Council offices and ask for a copy. Construct two different diagrams, each of which is appropriate for displaying the data.

3 How much pocket money do you receive per week? Give a breakdown of how you spend your pocket money and display it in a pie chart.

Obtain the same information from a friend. Draw another pie chart for your friend's data, choosing the radius of this circle so as to reflect any difference there might be between the two amounts of pocket money.

Review Exercises
Chapter 2

☐ LEVEL 1

1 Figure 2.10 relates to the most favoured leisure time activities of the pupils in a class.

Figure 2.10

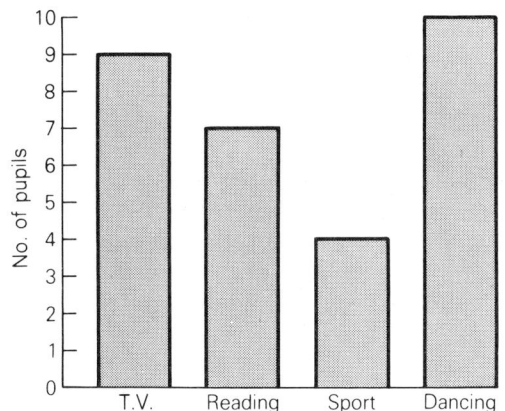

(a) How many pupils were there in the class?
(b) What is the modal activity?
(c) Draw a pie chart to illustrate the data.
(d) Draw a component bar diagram to illustrate the data.

2 Twenty cards were dealt one after another from an ordinary pack of playing cards and the suit of each card was noted. The results obtained were as follows: Heart, Club, Club, Spade, Diamond, Heart, Diamond, Heart, Club, Spade, Club, Diamond, Spade, Heart, Diamond, Spade, Club, Club, Heart, Heart.
(a) Construct a tally chart to determine the frequency of each suit.
(b) Draw a bar diagram to illustrate the results.
(c) Draw a pie chart to illustrate the results.

3 Table 2.13 shows a market research company's estimates of the total retail selling prices

Table 2.13

Type	RSP (£)
Computer games	100
Table-top games	40
Video consoles	25
Video games	15
Total	180

(RSPs), in millions of pounds, of various electronic games during 1985. Draw a pie chart for these data.

4 Table 2.14 shows some characteristics of the 30 houses on a housing estate. By first constructing tally charts, present the data on type of house and colour of front door as frequency distribution tables. Draw one kind of diagram for the distribution of the type of house and another kind of diagram for the colour of the front door; in each case name the kind of diagram you have drawn.

Table 2.14

No.	Type	Colour of front door	No.	Type	Colour of front door
1	Detached	Blue	16	Semi	Red
2	Detached	Brown	17	Detached	Brown
3	Semi	Red	18	Detached	Blue
4	Semi	Blue	19	Detached	Blue
5	Semi	Green	20	Detached	Green
6	Semi	Green	21	Semi	Yellow
7	Link	Brown	22	Semi	Green
8	Link	Blue	23	Semi	Yellow
9	Link	Green	24	Semi	Blue
10	Link	Blue	25	Semi	Brown
11	Link	Brown	26	Semi	Red
12	Link	Yellow	27	Detached	Blue
13	Semi	Blue	28	Detached	Green
14	Semi	Green	29	Detached	Red
15	Semi	Brown	30	Detached	Brown

☐ LEVEL 2

5 The areas of the oceans of the world expressed as percentages of the total ocean area are given in Table 2.15. Draw a component bar diagram to display these data.

Table 2.15

Ocean	Percentage of total area
Antarctic	5
Arctic	3
Atlantic	27
Indian	19
Pacific	46

6 A county reported that its gross expenditure during the last financial year totalled £286·5m, of which £168·4m was spent on Education, £28·4m on Social Services, £36·3m on Highways,

[15

£30.3m on Police, and the remainder on various other services. Prepare a table to display this information and exhibit it as (a) a component bar diagram, (b) a pie chart.

7 Table 2.16 shows the number of M.P.s of each political party that were elected in the election of June 1983. Draw a pie chart to display these results.

Table 2.16

Party	No. elected
Conservative	396
Labour	209
Liberal	17
SDP	6
Other	22
Total	650

8 In an opinion poll each of 640 male voters and 360 female voters was asked which political party he/she would vote for if there was a general election the following day. The responses obtained are shown in Table 2.17.

Table 2.17

Party	Males	Females	Total
Conservative	180	139	319
Labour	206	84	290
Alliance	98	54	152
Others	86	47	133
Don't know	70	36	106
Total	640	360	1000

(a) Draw a bar diagram to display the numbers in the last column (Total).
(b) Draw a pie chart to display the males' responses, using a circle of radius 3 cm.
(c) Draw a dual percentage bar diagram appropriate for a visual comparison of the voting breakdowns of males and females and comment on any differences.
(d) If the responses are to be compared by means of pie charts, with the males' responses being represented by a circle of radius 3 cm, calculate the radius of the circle that should be used to represent the females' responses.

3

SMALL NUMERICAL DATA SETS

"I got 100% in French and 0% in Maths – Does that make me half a genius or just O.K. on average?"

In this chapter we consider the presentation and analysis of a fairly small number of observations (say up to 20) of a discrete or continuous quantitative variable. Larger data sets will be covered in later chapters.

3.1 Diagrams

○ *EXAMPLE 3.1*

A householder's weekly consumptions of electricity in kilowatt-hours during a period of nine weeks in a winter were as follows:

338, 354, 341, 353, 351, 341, 353, 346, 341.

For a small numerical data set, one way of displaying the observations is by means of a **dot diagram**. The dot diagram for the electricity consumption data is shown in Figure 3.1. Here a horizontal line has been drawn and scaled to represent the variable values, and a dot (or circular blob) has been placed above the value of each observation.

Figure 3.1 *Dot Diagram for Example 3.1*

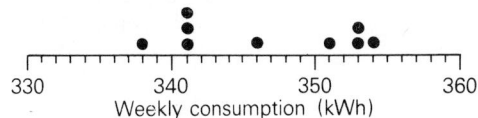
Weekly consumption (kWh)

Alternatively we may use a **stem and leaf diagram**. In this type of diagram we first draw a vertical line called the 'stem' (like the stem of a plant). To the left of the stem we write down the first figure or figures of the values of the observations that show variation.

[17

For the electricity consumption data the first figure in each observed value is 3, so there is no variation. Looking at the second figure there is variation, since it may be either 3, 4 or 5. So to the left of the stem we put in the figures 33, 34 and 35, as shown in Figure 3.2a. To the right of the stem we record the remaining figure in each observed value (the 'leaves'). This is shown in Figure 3.2b. Finally we order the figures in each row to the right of the stem by size to give the completed stem and leaf diagram shown in Figure 3.2c.

Figure 3.2 *Stem and Leaf Diagram for Example 3.1.*

```
(a) 33 |           (b) 33 | 8           (c) 33 | 8
    34 |               34 | 1 1 6 1         34 | 1 1 1 6
    35 |               35 | 4 3 1 3         35 | 1 3 3 4
                                          Stem   Leaves
```

○ **EXAMPLE 3.2**

The times, in minutes, taken by 20 people to complete a certain task were:

   35, 49, 41, 31, 35, 56, 36, 27, 45, 52,
   16, 42, 29, 51, 38, 31, 46, 53, 33, 32.

The dot diagram for this set of data is given in Figure 3.3.

Figure 3.3 *Dot Diagram for Example 3.2*

To draw the stem and leaf diagram we first note that there is variation in the first figure of the observed values; this may be 1, 2, 3, 4 or 5. So to the left of the stem we write in the figures from 1 to 5, and to the right the associated second figures. This step is shown in Figure 3.4a. Reordering the figures to the right of the stem by size gives us the diagram shown in Figure 3.4b.

Figure 3.4 *Stem and Leaf Diagram for Example 3.2*

```
(a) 1 | 6                (b) 1 | 6
    2 | 7 9                  2 | 7 9
    3 | 5 1 5 6 8 1 3 2      3 | 1 1 2 3 5 5 6 8
    4 | 9 1 5 2 6            4 | 1 2 5 6 9
    5 | 6 2 1 3              5 | 1 2 3 6
```

● **EXERCISE 3.1**

Draw a dot diagram and a stem and leaf diagram for each of the following sets of data:
1 23, 64, 58, 46, 51, 62, 51, 46, 38, 46.
2 57, 96, 84, 73, 92, 81, 65, 84, 86, 71, 84, 85.
3 124, 146, 135, 144, 142, 132, 135, 142, 128, 133, 151, 147.

3.2 Summary Measures of a Data Set

A **summary measure** is a quantity whose value measures some particular feature of a set of data. We consider two features. One feature of interest is the general magnitude of the observed values. A summary measure for this feature is known as a **measure of location**. Another feature of interest is the extent of the variation in the values of the observations. A summary measure for this feature is known as a **measure of dispersion**.

☐☐ MEASURES OF LOCATION

A **measure of location** for a numerical data set is a single value chosen to be representative of the values of the observations. It is sometimes referred to as an **average**, or a **measure of central tendency**. We shall refer to it as a measure of location since its value gives some indication of the location (whereabouts) of the observations on a scaled line. There are three commonly used measures of location, namely the **mode**, the **median**, and the **mean**.

☐ THE MODE

The **mode** of a data set is the value that occurs most frequently (if there is one). For the data in Example 3.1 the mode is 341, which occurs three times; no other value occurs that many times.

The mode is of very limited use with small data sets. This is because it often happens that a small data set has no mode. For instance, there is no mode for the data in Example 3.2. This is because each of the values 31 and 35 occurs twice, with no other value occurring more frequently. As another example, consider the data:

   338, 340, 341, 343, 344, 346.

All of these values are different, so that again there is

no mode. The mode is most relevant when dealing with large data sets, covered in the next chapter.

☐ THE MEAN

The **mean** (more strictly, the **arithmetic mean**) of a data set is simply the average (as used in everyday speech) of all the observations. Its value is calculated by finding the sum of all the observations and dividing the result by the number of observations. Writing x for the value of an observation, the mean is written as $\bar{x}$ (pronounced as 'x bar').

For the data in Example 3.1 the mean weekly consumption in kilowatt-hours is:

$$\bar{x} = \frac{338+354+341+353+351+341+353+346+341}{9}$$
$$= 346\tfrac{4}{9} \text{ kW h}.$$

● EXERCISE 3.2

(Keep your answers for use in later exercises.)

1 Determine the mode (when there is one) and the mean of each of the following sets of data:
(a) 5, 3, 6, 6, 6, 7, 5, 3, 4.
(b) 8, 11, 8, 8, 7, 10.
(c) 9, 12, 9, 11, 10, 9, 10.
(d) 9, 12, 9, 11, 10, 9, 31.
(e) 1·5, 3·3, 2·8, 2·5, 3·1, 1·5, 2·1, 3·2.
(f) 225, 235, 220, 250, 280.
(g) 25, 118, 125, 101, 116.

2 Find the value of x if the nine numbers: 103, 106, 105, 103, 104, 106, 107, 105, x, have mean 105. Write down the mode of the nine numbers.

☐ ADVANTAGES AND DISADVANTAGES OF THE MEAN

Some of the advantages of the mean as a measure of location of a set of data are as follows:
(1) It is fairly easy to calculate.
(2) It is the average value in the everyday meaning of the word 'average'.

EXAM FAILURES

The following are all examples of bad examination questions because they require candidates to do things which do not make sense statistically. Take a careful look at them and decide why they are not good questions.

1 The average weight of 7 ballerinas is 46 kg. The average weight of 4 rugby players is 92 kg. What is the average weight of all eleven people?

2 During the last five days, the 08.22 train has been late five times:

	No. of minutes late
Monday	22 (fire in restaurant car)
Tuesday	4
Wednesday	1
Thursday	5
Friday	2

How late is this train, on average?

If a candidate gave the answer '3 minutes' to this last question, why should it be preferred to the arithmetic mean of 6·8 minutes?

3 What is the mean of 4, 31, 32, 33, 31, 31?

Perhaps you could rewrite these three questions to improve them.

(3) It is a fairly central value in the following sense. In Figure 3.1, imagine the variable axis as being a light rod and the dots above the axis as being unit masses resting on the rod. Then the point at which the rod will balance (the centre of gravity of the masses) is at $\bar{x}$.

(4) It is by far the most important measure of location in an area of Statistics (beyond the level of this book) known as statistical inference. This arises when the data set we have is only a sample from a much larger data set and the aim is to use the sample data to say something about the mean of the larger data set.

Some disadvantages of the mean are as follows:

(1) It may not be an observed value, and as such may not always be a good choice of a single value representation of the observations. For instance, we showed that the mean of the data in Example 3.1 was $346\frac{4}{9}$, which is not one of the observed values.

(2) Its value is affected by the presence of an observation which is much greater or much smaller than the other observations. To see this, suppose that in Example 3.1 the ninth week of the period had been very cold and that the consumption during that week had been 454 kW h instead of 341 kW h. Then the mean consumption over the nine weeks becomes 359 kW h. Since this is bigger than all but one of the nine observations, it is hardly a good choice as a single value representation of weekly consumptions.

Questions *3–9* *For each of the data sets (a)–(g) in Question 1, Exercise 3.2 state whether or not the mean you calculated is a good choice of a single value to represent the observations.*

☐ THE MEDIAN

The **median** of a data set is the centrally placed value when the observations are ordered from the smallest to the largest (or from the largest to the smallest). We shall use the letter m to denote the median of a data set.

Ordering the data of Example 3.1 from smallest to largest we have:

338, 341, 341, 341, 346, 351, 353, 353, 354.

Since there are nine observations, the central one is the fifth, so that the median is $m=346$. In this example the median does not differ very much from the mean $\bar{x}=346\frac{4}{9}$.

A problem arises when the number of observations is even. This is because there will be no central observation. The convention then is to take the median to be the average of the *two* central observations. To illustrate this, suppose that in Example 3.1 the consumption in the tenth week was 355. Adding this to the data of Example 3.1 and ordering the ten observations from smallest to largest we get:

338, 341, 341, 341, 346, 351, 353, 353, 354, 355.

The two centrally placed observations are the 5th (346) and the 6th (351). The median now is:

$$m = \tfrac{1}{2}(346+351) = 348\tfrac{1}{2}.$$

For a data set of n observations ordered from smallest to largest:

(a) when n is odd the median is the value of the $\tfrac{1}{2}(n+1)$th observation,
(b) when n is even the median is the average of the $(\tfrac{1}{2}n)$th and the $(\tfrac{1}{2}n+1)$th observations.

☐ ADVANTAGES AND DISADVANTAGES OF THE MEDIAN

Some advantages of the median as a measure of location are as follows:

(1) It is easy to determine when the observations have been ordered.
(2) It is, by definition, a central value.
(3) Unlike the mean, the presence of an abnormally small or large observation has no effect on its value. For example, the median in Example 3.1, is still 346, even if the largest observation of 354 is increased to some larger value or if the smallest observation 338 is decreased to some smaller value, or if both apply.
(4) For an odd number of observations the median will always be one of the observed values (the middle one). As such, the median is often more appropriate than the mean as a single value representation of the data. This may or may not be the case for an even number of observations. For an even number of observations the median will be an observed value only if the two central observations happen to have the same value.
(5) The fact that only the central value (or values) need be known for the median can be very advantageous in some cases. In particular, to determine the median height of the girls in a class we may line up the girls from the shortest to the tallest and merely measure the height of the girl (or girls) in the middle of the line.

The only serious disadvantage of the median is in statistical inference (referred to earlier) in which the mean turns out to be the most useful measure of location.

Questions **10–16** *Calculate the median of each of the data sets (a)–(g) in Question 1, Exercise 3.2. Keep your answers to these questions for use in Questions 29–35, Exercise 3.3.*

☐☐ MEASURES OF DISPERSION OR SPREAD

Having determined a single value (a measure of location) as being representative of a data set, another feature that is usually of interest is the extent of the variation among the observations. A measure of this variation is referred to as a **measure of dispersion** or a **measure of spread**, or occasionally as a **measure of variation**. Here we shall discuss four measures of dispersion that are appropriate for small data sets.

☐ THE RANGE

The **range** of a data set is the difference between the values of the smallest and largest observations. For Example 3.1 the range of the weekly consumptions of electricity is $354-338=16$ kWh.

Since the range depends only on the two most extreme values it is very sensitive to an abnormally small or large value in a data set. As such, it is appropriate only when dealing with a small data set. The larger the data set the less informative is the range as a measure of the variation in the values of the observations.

☐ MEAN ABSOLUTE DEVIATIONS

To get a measure of variation it would be sensible to look at the deviations of the values of the observations from some central value. Either the median or the mean is appropriate. Combining such deviations in some way will give a measure of dispersion. Taking their average is one possibility. But since the deviations are from a central value some will be positive, some will be negative, and maybe some will be zero. So the average could be very small (or even zero) even when there is considerable variation in the observations. For this reason it is better to use **absolute deviations**, which are the numerical values (ignoring minus signs) of the actual deviations. An absolute deviation cannot be negative. Averaging the absolute deviations gives a measure of dispersion known as the **mean absolute deviation**, which we abbreviate as MAD. Since either the median or the mean may be chosen as the central value there are two such measures, the **mean absolute deviation from the mean** and the **mean absolute deviation from the median**.

☐ THE MEAN ABSOLUTE DEVIATION FROM THE MEAN

When using $\bar{x}$ as the measure of location it is logical to average the absolute deviations from $\bar{x}$ for an associated measure of dispersion. This measure is called the mean absolute deviation from the mean, which we abbreviate as MAD($\bar{x}$). We have

$$\text{MAD}(\bar{x}) = \frac{\text{sum of absolute deviations from } \bar{x}}{\text{number of observations}}$$

○ EXAMPLE 3.3

Find the mean absolute deviation from the mean for the data in Example 3.1.

The data are:

338, 354, 341, 353, 351, 341, 353, 346, 341.

We showed earlier that $\bar{x} = 346\frac{4}{9}$. The deviations from this value (that is, the values of $x - \bar{x}$ for each observation) are:

$-8\frac{4}{9}, 7\frac{5}{9}, -5\frac{4}{9}, +6\frac{5}{9}, 4\frac{5}{9}, -5\frac{4}{9}, 6\frac{5}{9}, -\frac{4}{9}, -5\frac{4}{9}$.

The sum of these deviations is zero. We made the point earlier that averaging the deviations can lead to an unsatisfactory measure of dispersion. (In fact it is always the case that for any data set the average of the actual deviations from the mean will be zero.)

Ignoring the minus signs, the sum of the absolute deviations is:

$8\frac{4}{9} + 7\frac{5}{9} + 5\frac{4}{9} + 6\frac{5}{9} + 4\frac{5}{9} + 5\frac{4}{9} + 6\frac{5}{9} + \frac{4}{9} + 5\frac{4}{9} = 50\frac{4}{9}$.

The mean absolute deviation from the mean is:

$$\text{MAD}(\bar{x}) = 50\frac{4}{9}/9 = 454/81 = 5\cdot60 \text{ (to 2 d.p.)}.$$

If a calculator is used to evaluate MAD($\bar{x}$) in the above example it is tempting to write $\bar{x}$ in decimal form before calculating the absolute deviations and summing them. This is not advisable in our example because decimalising $\bar{x}$ will produce a recurring decimal part. This will have to be rounded when it is entered into the calculator, the number of figures kept being dependent on the capacity of the calculator. By rounding one introduces a small error in each absolute deviation, and the answer for MAD($\bar{x}$) will be slightly in error.

THE MEAN ABSOLUTE DEVIATION FROM THE MEDIAN

When the median is used as a measure of location the mean absolute deviation from the median, MAD(m) is an appropriate associated measure of dispersion. It is given by:

$$\text{MAD}(m) = \frac{\text{sum of absolute deviations from } m}{\text{number of observations}}.$$

EXAMPLE 3.4

Find the mean absolute deviation from the median for the data in Example 3.1.

Ordering the data in Example 3.1 from the smallest to the largest we have:

338, 341, 341, 341, 346, 351, 353, 353, 354.

Since $m = 346$ (shown earlier),

$$\text{MAD}(m) = \frac{8+5+5+5+0+5+7+7+8}{9}$$
$$= \frac{50}{9} = 5\tfrac{5}{9} = 5\cdot 56 \quad \text{(to 2 d.p.).}$$

In our example it is clear that m and MAD(m) are easier to calculate than $\bar{x}$ and MAD($\bar{x}$), and this will often be the case.

Two advantages of both MADs are that they are fairly easy to calculate, and that, being averages of deviations, they can be seen to be measures of the dispersion (or spread) of the observations. A disadvantage is that their values will be magnified if the data set includes an abnormally small or large value. (To see this, compare your answers to Questions 3 and 4 of the following exercise.) The main disadvantage of both MADs is that they are unsuitable for use in statistical inference.

EXERCISE 3.3

Questions **1–7** *Find the range, the mean absolute deviation from the mean, and the mean absolute deviation from the median, of each of the data sets (a)–(g) in Question 1, Exercise 3.2.*

THE STANDARD DEVIATION

Another way of combining the deviations from $\bar{x}$ to get a measure of dispersion is to take the average of their squares. The result is called the **variance** of the data set, abbreviated as VAR. This is an alternative to MAD($\bar{x}$) as a measure of dispersion to use in conjunction with the mean $\bar{x}$ as the measure of location. However, since the deviations have been squared the units of VAR will be the square of the units of the variable. In Example 3.1 the observations are in kilowatt-hours, so that the units of the variance will be (kilowatt-hours)^2. For this reason it is more usual to take the square root of the variance as the measure of dispersion. The resulting quantity is called the **standard deviation**, abbreviated SD, and it will have the same units as the variable. The expressions for the variance and standard deviation are:

$$\text{VAR} = \frac{\text{sum of squares of deviations from } \bar{x}}{\text{number of observations}},$$

$$\text{SD} = \sqrt{\text{VAR}}.$$

EXAMPLE 3.5

Find the standard deviation of the data in Example 3.1.

The data of Example 3.1 are:

338, 354, 341, 353, 351, 341, 353, 346, 341.

From Example 3.3, the deviations of these observations from their mean $\bar{x} = 346\tfrac{4}{9}$ are:

$-8\tfrac{4}{9}, 7\tfrac{5}{9}, -5\tfrac{4}{9}, 6\tfrac{5}{9}, 4\tfrac{5}{9}, -5\tfrac{4}{9}, 6\tfrac{5}{9}, -\tfrac{4}{9}, -5\tfrac{4}{9}.$

Writing these as integers over 9, which is more convenient when using a calculator, they become:

$-\tfrac{76}{9}, \tfrac{68}{9}, -\tfrac{49}{9}, \tfrac{59}{9}, \tfrac{41}{9}, -\tfrac{49}{9}, \tfrac{59}{9}, -\tfrac{4}{9}, -\tfrac{49}{9}.$

The sum of the squares of these deviations is:

$$\frac{76^2 + 68^2 + 49^2 + 59^2 + 41^2 + 49^2 + 59^2 + 4^2 + 49^2}{81}$$

$$= \frac{26\,262}{81}.$$

The variance of the data in Example 3.1 is then:

$$\text{VAR} = \left(\frac{26\,262}{81}\right)/9 = \frac{26\,262}{729} \ (\text{kWh})^2.$$

Taking the square root, the standard deviation is:

$$\text{SD} = \sqrt{\frac{26\,262}{729}} = 6.00 \text{ kWh},$$

to two decimal places.

The disadvantages of the standard deviation are:
(1) Its value is magnified when there is an abnormally small or large observation in the data set.
(2) It is more difficult than any of the other measures to see as being a measure of dispersion.
(3) It is fairly complicated to calculate as compared with the other measures of dispersion.

Despite these disadvantages the standard deviation is by far the most important measure of dispersion in advanced studies in Statistics.

☐ USING CALCULATORS TO FIND THE MEAN AND STANDARD DEVIATION

Some care is needed when using a calculator which is preprogrammed to give the mean and the standard deviation of a data set. With such a calculator each observation is entered and a special key, often labelled x_n, is pressed; this is done for all the observations. Having entered all the observations, a key usually labelled $\bar{x}$ is pressed for the value of the mean to be displayed. Another key labelled σ_n or σ_{n-1} or s (some calculators have two such keys) is then pressed to display the standard deviation. If the key is labelled σ_n the value displayed will be the standard deviation as defined here. The subscript n is the number of observations in the data set and is the divisor we used in calculating the variance. If the key is labelled σ_{n-1} or s the value displayed is that obtained when the divisor n is replaced by $n-1$. (The use of the divisor $n-1$ has a justification in the area of statistical inference but is not appropriate in descriptive or summary statistics.) If you are in doubt about your calculator refer to the manual.

Because of the limited storage space in a calculator problems can arise when the observations are very large but do not vary very much. Consider the data:

1 000 000, 1 000 001, 1 000 002, 1 000 003.

The values here are very large but there is little variation between them. If you have a calculator with a preprogrammed standard deviation facility, enter the above values and try to find their standard deviation. You will probably obtain the answer zero. This is clearly wrong since there is some variation in the values of the data set. A way of overcoming this is given in Section 3.3.

Questions **8–14** *Calculate, to two decimal places, the standard deviation of each of the data sets (a)–(g) in Question 1, Exercise 3.2.*

☐ ALTERNATIVE EXPRESSION FOR SD

The calculation of a standard deviation is often easier using an alternative formula as follows. It is shown in the box on p. 32 that:
The sum of the squares of the deviations from $\bar{x}$

= (sum of squares of observations)
$- \left(\dfrac{\text{square of sum of observations}}{\text{number of observations}} \right)$

This can be expressed more simply in symbols by using the Greek capital letter Σ (pronounced 'sigma')
to mean 'the sum of'. If x is an arbitrary observation in the data set then Σx is the sum of all the observations, and $\Sigma(x-\bar{x})^2$ is the sum of the squares of the deviations of the x-values from $\bar{x}$. For n observations, the expression in words above may then be written as:

$$\Sigma(x-\bar{x})^2 = \Sigma x^2 - \frac{1}{n}(\Sigma x)^2 \qquad (3\cdot1)$$

For the data in Example 3.5:

$\Sigma x = 338 + 354 + \ldots + 346 + 341 = 3118,$
$\Sigma x^2 = 338^2 + \ldots + 341^2 = 1\,080\,538.$

Using Expression (3.1) we have:

$$\Sigma(x-\bar{x})^2 = 1\,080\,538 - \frac{3118^2}{9} = \frac{2918}{9} = \frac{26\,262}{81},$$

as obtained earlier.
In symbols,

$$\text{VAR} = \frac{1}{n}\Sigma(x-\bar{x})^2 = \frac{1}{n}\left[\Sigma x^2 - \frac{1}{n}(\Sigma x)^2\right]$$
$$= \frac{1}{n^2}\left[n\Sigma x^2 - (\Sigma x)^2\right],$$

and:

$$\text{SD} = \frac{1}{n}\sqrt{\left[n\Sigma x^2 - (\Sigma x)^2\right]} \qquad (3\cdot2)$$

Using expression (3.2) for the data in Example 3.5, and the values of Σx and Σx^2 given above we find:

$$\text{SD} = \frac{\sqrt{((9 \times 108\,538) - 3118^2)}}{9} = \frac{\sqrt{2918}}{9}$$
$$= 6\cdot00 \quad \text{(to 2 d.p.)},$$

exactly as obtained earlier.

Using Expression (3.2) with a calculator is particularly advantageous, since both the sum (Σx) and the sum of squares (Σx^2) of a data set can be obtained on most calculators from a single entry of the observations. This is possible on any calculator that has a squaring key (usually labelled x^2) and an accumulating memory (usually labelled M+ or SUM or Σ). Since some calculators do not clear their stores when switched off you should first check that the contents of the stores are zero before using the following procedure.

To obtain the values of Σx and Σx^2 on a calculator the procedure is shown schematically in Figure 3.5.

Figure 3.5

Continue with this procedure for all the observations. The display will then show the sum of the squares (Σx^2) of all the observations. Make a note of its value or transfer it into a store (usually labelled STO) if you have one on your calculator as well as the accumulating memory. Now press the recall key for what is in store in the accumulating memory (this key is usually labelled RCL and may have an identification number if your calculator has more than one memory store). The number now shown on the display is the sum (Σx) of all the observations.

Calculators having the x_n key, and possibly one or more of the σ_n, σ_{n-1} or s keys described earlier, will also have keys which will give the sum (usually labelled Σx) and the sum of squares (usually labelled Σx^2).

Questions **15–21** *Use the alternative method (Expression (3.2)) given in the previous section to find, to two decimal places, the standard deviation of each of the data sets (a)-(g) in Question 1, Exercise 3.2. Check your answers against those you obtained in Questions 8-14.*

☐ **THE SEMI-INTERQUARTILE RANGE**

The median m of a data set is that value which divides the ordered observations into two halves. Now consider dividing the ordered observations into four quarters. The three values required to do so are called the **quartiles** of the data set. The middle one of these three values is the median. The smallest one is called the **lower quartile** and the largest is called the **upper quartile**, which we abbreviate as LQ and UQ, respectively.

The difference, UQ–LQ, is called the **interquartile range**, while $\frac{1}{2}$(UQ–LQ) is called the **semi-interquartile range**. Each of these is appropriate as a measure of dispersion in association with the median as the measure of location. Here we shall use the semi-interquartile range, abbreviating it as SIQR.

One difficulty with quartiles is that one cannot find three values that will divide a data set into four *exactly* equal quarters. For this reason different definitions of the quartiles have been suggested, all aimed at dividing a data set into four *roughly* equal quarters. These various definitions will sometimes lead to different answers for the lower and upper quartiles. Here we consider one definition which has some advantages. A graphical method for finding quartiles, which some readers may find easier than the method given here, is described in Section 4.3, p. 39.

To find the quartiles of n observations ordered from the smallest to the largest, first divide n by 4. The result may be a whole number or a whole number plus a remainder which may be 1 or 2 or 3.

(a) Suppose that on dividing n by 4 the result is a whole number k plus a remainder. Then the quartiles are defined as follows:
  LQ is the value of the $(k+1)$th observation,
  UQ is the value of the $(n-k)$th observation.

(b) Now suppose that $n/4=k$, a whole number. Then the quartiles are defined as follows:
  LQ is the average of the values of the kth and $(k+1)$th observations,
  UQ is the average of the values of the $(n-k)$th and $(n-k+1)$th observations.

The following examples illustrate the procedures.

○ **EXAMPLE 3.6**

Find the quartiles of the data in Example 3.1.
Ordering the data of Example 3.1 we have:

338, 341, 341, 341, 346, 351, 353, 353, 354.

We showed earlier that the middle quartile, or median, is 346. For this data set $n/4 = 9/4 = 2 + \frac{1}{4}$. From (a) above, with $k=2$ and $n-k=7$, the quartiles are found as follows. LQ is the value of the 3rd observation:

$$LQ = 341.$$

UQ is the value of the 7th observation:

$$UQ = 353.$$

(Equivalently UQ is the value of the 3rd observation counting down from the highest observation.)

It follows that the semi-interquartile range of the data is:

$$SIQR = \tfrac{1}{2}(353 - 341) = 6.$$

○ **EXAMPLE 3.7**

Find the quartiles of each of the data sets:

(a) 21, 22, 22, 24, 25, 27, 28, 28
(b) 5, 6, 6, 7, 7, 7, 8, 10, 11, 11, 17

(a) Here $n=8$, so that the middle quartile or median m is the average of the values of the 4th and 5th observations:

$$m = \tfrac{1}{2}(24+25) = 24\tfrac{1}{2}.$$

Since $n/4=2$, using (b) above with $k=2$ and $n-k=6$, the quartiles are as follows. LQ is the average of the values of the 2nd and 3rd observations:

$$LQ = \tfrac{1}{2}(22+22) = 22.$$

UQ is the average of the values of the 6th and 7th observations:

$$UQ = \tfrac{1}{2}(27 + 28) = 27\tfrac{1}{2}.$$

It follows that the semi-interquartile range is:

$$SIQR = \tfrac{1}{2}(27\tfrac{1}{2} - 22) = 2\tfrac{3}{4}.$$

(b) Here $n = 11$, so that the middle quartile or median m is the value of the 6th observation:

$$m = 7.$$

Since $n/4 = 2 + \tfrac{3}{4}$, using (a) above with $k = 2$ and $n - k = 9$, the lower and upper quartiles are found as follows. LQ is the value of the 3rd observation:

$$LQ = 6,$$

UQ is the value of the 9th observation:

$$UQ = 11$$

and the semi-interquartile range is:

$$SIQR = \tfrac{1}{2}(11 - 6) = 2\tfrac{1}{2}.$$

An important advantage of the SIQR as a measure of dispersion as compared with the other measures considered is that it is not affected by one of the observations being abnormally small or large. Since this is also true for the median, it is recommended that when a data set includes an abnormally small or large value (as compared with the other values), the median should be used as the measure of location and the semi-interquartile range as the measure of dispersion.

Questions **22–28** Determine the semi-interquartile range for each of the data sets (a)–(g) in Question 1, Exercise 3.2. (Keep your answers for use in Questions 29–35, Exercise 3.3.)

☐☐ BOX-AND-WHISKER PLOTS

Some of the main features of a data set may be summarized diagrammatically by means of a **box-and-whisker plot**. Such a diagram shows the values of the smallest and largest observations, the median, and the quartiles.

The box-and-whisker plots for the two data sets in Example 3.7 are given in Figures 3.6a and b, respectively. In each of these diagrams a scaled line is first drawn to represent the variable values. The line extends from the smallest value to the largest value in the data set; these are indicated in the diagram by circular blobs. A rectangular box is drawn having its vertical sides at the values of the quartiles; the position of the median is shown by the vertical line within the box. The parts of the scaled line coming out of the box are the whiskers. Observe that the box covers the middle 50% of the observations in the data

Figure 3.6 *Box-and-Whisker Plots: (a) for the Data of Example 3.7a (b for the Data of Example 3.7b*

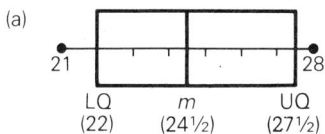
(a)

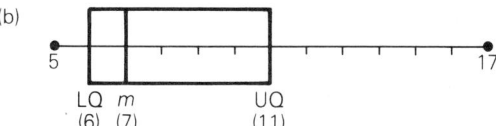
(b)

set. In Figure 3.6b the smallest value, the median and the lower quartile are fairly close to each other, which indicates that the observations are bunched at the bottom end of the scale. The long whisker to the right in Figure 3.6b shows that there are very few large observations.

Questions **29–35** Draw a box-and-whisker plot for each of the data sets (a)–(g) in Question 1, Exercise 3.2. (Use the answers you obtained in Questions 10–16, Exercise 3.2 and to Questions 22–28, Exercise 3.3)

3.3 Working Origins and Scales (Transformations)*

Sometimes the values in a data set can be simplified in a way which makes it easier to calculate summary measures.

○ EXAMPLE 3.8

The air pressures, in millibars, on eight successive days at a particular location were:

1025, 1030, 1025, 1010, 1015, 1010, 1005, 1000.

Since the values here are fairly large, we are more likely to make calculator entry and arithmetical errors. Notice that all the values are 1000 or more. If we subtract 1000 from each value we have the simpler numbers:

25, 30, 25, 10, 15, 10, 5, 0.

Figure 3.7 *Dot Diagrams*

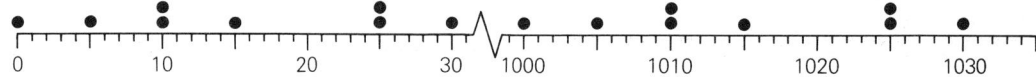

Now look at the dot diagrams of the two data sets, given in Figure 3.7. We see that the location of the new data set is simply a shift of the original data set by 1000 to the left. It follows that a measure of location for the original data set will have a value of 1000 more than that for the new data set. So if we calculate the median and the mean of the new data set, adding 1000 to each answer will give the median and the mean of the original data set.

Ordering the new data set from the smallest to the largest we have:

$$0, 5, 10, 10, 15, 25, 25, 30.$$

Since there are eight observations the median is the average of the fourth and fifth observations and is equal to $\frac{1}{2}(10+15) = 12 \cdot 5$. It follows that the median of the original data set is $1000 + 12 \cdot 5 = 1012 \cdot 5$, as you may easily check directly from the original data set.

The mean of the new data set is

$$\frac{0+5+10+10+15+25+25+30}{8} = \frac{120}{8} = 15,$$

so that the mean of the original data set is $1000 + 15 = 1015$, which you can again check directly from the original data set.

Another observation to make from Figure 3.7 is that the two sets of dots are identical apart from their location. This means that the two data sets must have equal dispersions (variations). So any measure of dispersion will have the same value for both data sets. Working on the new data set we find that:

the mean absolute deviation from the median is 8·75,
the mean absolute deviation from the mean is 8·75,
the standard deviation is 10,
the semi-interquartile range is 8·75.

These are also precisely the values for the original data set.

In general, suppose we have a data set of x-values. Now subtract some number c from each x-value. Taking $y = x - c$ for each x-value we get a new data set, of y-values. The value c is referred to as the **working origin**, and the relation $y = x - c$ is referred to as a **transformation** from the x-values to the y-values. Using an obvious notation, the summary measures for the data set of x-values can be obtained from those of the data set of y-values as follows:

$$\begin{aligned} m_x &= m_y + c, \\ \bar{x} &= \bar{y} + c, \\ \text{MAD}(m_x) &= \text{MAD}(m_y), \\ \text{MAD}(\bar{x}) &= \text{MAD}(\bar{y}), \\ \text{SIQR}(x) &= \text{SIQR}(y). \\ \text{SD}(x) &= \text{SD}(y). \end{aligned} \quad (3.3)$$

● EXERCISE 3.4

1 Work directly on the original data set (1025, 1030, etc.) in Example 3.8 to check the values of the summary measures obtained above.

2 Subtract 200 from each of the observations in data set (f) of Question 1, Exercise 3.2. Calculate the values of the various summary measures for this new data set. Deduce the corresponding values for the original data set. Check the answers you get with those you obtained in earlier exercises.

3 Add 10 to each observation in data set (ᴧ) of Question 1, Exercise 3.2. Calculate the values of the various summary measures for this new data set. Deduce the corresponding values for the original data, checking them with the answers you obtained in earlier exercises.

☐☐ USING A WORKING SCALE

Looking again at the data set in Example 3.8 we see that each value is an exact multiple of 5. Dividing each value by 5 gives the simpler set of numbers:

$$205, 206, 205, 202, 203, 202, 201, 200.$$

By dividing the values by 5 we are changing the **scale of measurement**. A difference of 5 units in two values in the original data set becomes a difference of 1 unit in the corresponding values in the new data set. This means any measure of location of the original data set will have a value which is five times that of the new data set. For the new data set we find that the median is 202·5 and the mean is 203. To obtain the median and the mean of the original data we need only multiply these answers by 5 to find that the median is 1012·5 and the mean is 1015, agreeing with the answers we obtained earlier.

The dot diagrams for the original and new data sets are shown in Figure 3.8. We see that the

Figure 3.8 *Dot Diagrams*

observations in the new data set are more tightly bunched than those in the original data set. This means that the dispersion of the values in the original set is greater than that in the new data set. In fact we shall show that each measure of dispersion for the original data set is exactly five times that for the new data set. Letting y represent an observation in the new data set it may be shown that:

$$\text{MAD}(m_y) = 1 \cdot 75,$$
$$\text{MAD}(\bar{y}) = 1 \cdot 75,$$
$$\text{SD}(y) = 2,$$
$$\text{SIQR}(y) = 1 \cdot 75.$$

Referring back to the values we got for the original data set we see that they are exactly 5 times those given above.

The transformation we used in this example was $y = x/5$. In general, for the transformation $y = x/k$, for any chosen k, *each summary measure of the x-values has a value which is exactly k times that of the y-values.*

4 *Divide by 5 each of the observations in data set (f) of Question 1, Exercise 3.2. Calculate the values of the summary measures for this new data set. Deduce the corresponding values for the original data set, checking your answers with those you obtained in earlier exercises.*

5 *Multiply by 10 each observation in data set (a) of Question 1, Exercise 3.1. Calculate the values of the summary measures for the data set you have generated. Deduce the corresponding values for the original data set, checking your answers with those you obtained in earlier exercises.*

☐☐ USING A WORKING ORIGIN AND SCALE TOGETHER

We have seen how the calculation of summary measures can be simplified by using a working origin and by using a working scale. The calculation can sometimes be made simpler still by combining the two.

The first thing we did with the data in Example 3.8 was subtract 1000 from each value to get the new data:

25, 30, 25, 10, 15, 10, 5, 0. data set 1

Notice that each of these numbers is an exact multiple of 5. So now let us divide each value by 5 to end up with the very simple numbers:

5, 6, 5, 2, 3, 2, 1, 0. data set 2

The transformation we have used from the original values (x) to get the new values (y) is:

$$y = (x - 1000)/5.$$

First consider the mean. We know that the mean of data set 1 will be 5 times the mean of data set 2. We also know that the mean of the original data set (1025, 1030, etc.) will be 1000 more than the mean of data set 1. It is easily shown that the mean of data set 2 is $\bar{y} = 3$. It follows that the mean of data set 1 is $3 \times 5 = 15$, and therefore the mean of the original data set is $1000 + 15 = 1015$, as shown earlier.

The median of the original data set can be obtained similarly from the median of data set 2 and is equal to $1000 + (5 \times 2 \cdot 5) = 1012 \cdot 5$.

Now consider the standard deviation. We know that the standard deviation of data set 1 is 5 times that of data set 2. We also know that the standard deviation of the original data set is equal to that of data set 1. It may be shown that the standard deviation of data set 2 is $\text{SD}(y) = 2$. It follows that the standard deviation of data set 1 is $5 \times 2 = 10$, and therefore the standard deviation of the original data set is 10, as shown earlier.

In general, if we use the transformation:

$$y = (x - c)/k,$$

then:
$$\bar{x} = k\bar{y} + c$$
$$m_x = km_y + c,$$
$$\text{SD}(x) = k\ \text{SD}(y),$$
$$\text{MAD}(m_x) = k\ \text{MAD}(m_y), \qquad (3.4)$$
$$\text{SIQR}(x) = k\ \text{SIQR}(y).$$

○ **EXAMPLE 3.9**

Use an appropriate transformation to find the mean and the standard deviation of the data set:

11·3, 12·2, 14·1, 13·3, 13·4.

Let us first subtract 11 from each observation to get the data set:

0·3, 1·2, 3·1, 2·3, 2·4.

Now get rid of the decimals by multiplying by 10.

This gives the data set of simple numbers.

$$3, 12, 31, 23, 24.$$

The transformation we have used here is:

$$y = (x-11) \times 10,$$

so that $c=11$ and $k=\frac{1}{10}$. For the transformed data set (of y-values) we find:

$$\Sigma y = 3 + 12 + 31 + 23 + 24 = 93,$$
$$\Sigma y^2 = 3^2 + 12^2 + 31^2 + 23^2 + 24^2 = 2219.$$

It follows that:

$$\bar{y} = \frac{1}{n}\Sigma y = \frac{93}{5} = 18 \cdot 6,$$

and

$$SD(y) = \frac{1}{n}\sqrt{(n\Sigma y^2 - (\Sigma y)^2)}$$
$$= \frac{\sqrt{(5 \times 2219 - 93^2)}}{5} = \frac{\sqrt{2446}}{5}.$$

Using Expressions (3.4) with $c=11$ and $k=0 \cdot 1$, the mean and the standard deviation of the original data set of x-values are:

$$\bar{x} = k\bar{y} + c = 0 \cdot 1 \bar{y} + 11 = 1 \cdot 86 + 11 = 12 \cdot 86,$$

and

$$SD(x) = k\ SD(y) = 0 \cdot 1\ SD(y) = \frac{0 \cdot 1 \sqrt{2446}}{5} = 0 \cdot 99,$$

to two decimal places.

6 Apply the transformation $y=(x-220)/5$ to data set (f) in Question 1, Exercise 3.2. For the transformed data set calculate the mean, the median, the mean absolute deviations from the median and from the mean, the standard deviation, and the semi-interquartile range. Deduce the corresponding values for the original data set, checking your answers with those you obtained in earlier exercises.

7 Use an appropriate transformation in each case to calculate the mean and the standard deviation for each of the following data sets:

(a) 5200, 5300, 5500, 5100, 5200.
(b) 1·65, 3·05, 2·15, 2·75, 3·15, 2·45, 1·85, 1·95.

☐☐ TRANSFORMING TO A GIVEN MEAN AND STANDARD DEVIATION*

A data set of x-values has mean $\bar{x}$ and standard deviation $SD(x)$. Suppose that for some reason we want to convert the set of x-values into a new set of values whose mean and standard deviation have been specified. The procedure for doing this is illustrated in the following example. An application of this procedure is given in the exam mark example on p. 29.

○ EXAMPLE 3.10

Consider again the data in Example 3.8:

$$1025, 1030, 1025, 1010, 1015, 1010, 1005, 1000.$$

Convert this set of values into a set of new values

(a) having mean 0 and standard deviation 1,
(b) having mean 50 and standard deviation 5.

From Example 3·8, we know that the above have mean $\bar{x} = 1015$, and standard deviation $SD(x) = 10$.

(a) From page 26, we know that subtracting $\bar{x}$ from all the values will give a set of values having mean 0. Subtracting 1015 from each of the values given we have

$$10, 15, 10, -5, 0, -5, -10, -15, \qquad (1)$$

whose mean is zero. We also know that this operation will not change the value of the standard deviation, so that the standard deviation of the set of values (1) is also 10.

Again from page 27, we know that in order to reduce the standard deviation from 10 to 1 we need to divide all the values in (1) by 10. Denote such a value by z. The z-values are:

$$1, 1 \cdot 5, 1, -0 \cdot 5, 0, -0 \cdot 5, -1, -1 \cdot 5. \qquad (2)$$

(Note that dividing values in a data set by 10 will also result in the new mean being the old mean divided by 10; but since the mean of (1) is 0 it follows that the mean of (2) is also 0. In Question 8 below you are asked to verify that the data set (2) really does have mean 0 and standard deviation 1.)

In general, the transformation to use to convert a set of x-values having mean $\bar{x}$ and standard deviation $SD(x)$ into a set of z-values having mean 0 and standard deviation 1 is given by

$$z = \frac{x - \bar{x}}{SD(x)} \qquad (3)$$

The resulting z-values are often referred to as the **standardized scores** of the x-values.

(b) Here, the easiest way of proceeding is to first convert the given x-values to z-values (standardized scores), which we have done in (a). We then convert the z-values into y-values having mean 50 and standard deviation 5. It turns out to be slightly easier to first adjust the standard deviation (unlike in (i) where we first adjusted the mean).

The z-values in (2) above have standard deviation 1. For a set of values having standard deviation 5 we need to multiply the values in (2) by 5. This gives the values

$$5,\ 7{\cdot}5,\ 5,\ -2{\cdot}5,\ 0,\ -2{\cdot}5,\ -5,\ -7{\cdot}5 \quad (4)$$

Since the mean of the z-values is 0, multiplying by 5 still gives the value 0 as the mean of the values in (4).

Let us now adjust the mean, which has to be increased from 0 to 50. Adding 50 to the values in (4) gives the values

$$55,\ 57{\cdot}5,\ 55,\ 47{\cdot}5,\ 50,\ 47{\cdot}5,\ 45,\ 42{\cdot}5. \quad (5)$$

Since adding 50 does not change the standard deviation these are the required y-values having mean 50 and standard deviation 5. You are asked to verify this in Question 8 below.

In general, to convert z-values (standardized scores) into y-values having mean $\bar{y}$ and standard deviation $SD(y)$, the steps are as follows:

Step 1: Multiply the z-values by $SD(y)$, so that z becomes $zSD(y)$.

Step 2: Add on $\bar{y}$ to the new values to end up with y-values, where for any z-value.

$$y = \bar{y} + zSD(y).$$

This is the transformation to use to convert z-values into y-values having mean $\bar{y}$ and standard deviation $SD(y)$.

Combining this transformation with (3), the transformation required to convert the x-values directly into the y-values is given by

$$y = \bar{y} + \left\{\frac{x - \bar{x}}{SD(x)}\right\} SD(y) \quad (6)$$

8 Verify that (a) the data set (2) in Example 3.10 has mean 0 and standard deviation 1, (b) the data set (5) in Example 3.10 has mean 50 and standard deviation 5.

9 Show that the data set

$$24,\ 16,\ 21,\ 38,\ 29,\ 28,\ 9,\ 20,\ 11,\ 34,$$

has mean 23 and standard deviation 9.
(a) Convert these values into (i) standardized scores, (ii) a set of values having mean 20 and standard deviation 1.8.
(b) Find the transformation that will convert the given data set into one having mean 50 and standard deviation 18.

☐☐ COMPARING EXAMINATION MARKS FOR DIFFERENT SUBJECTS*

How do you view examination marks? In particular, suppose that you got 66% in a Mathematics examination and that Moira, another pupil in your class, got 66% in an English examination. Does this mean that you are as good in Mathematics as Moira is in English? This question cannot be answered. But consider the related question: relative to the class as a whole is your Mathematics mark as good as Moira's English mark? This question can be answered provided that we know the means and the standard deviations of the marks for the whole class in the two examinations.

Suppose that the marks for the entire class in Mathematics and English are as shown in Table 3.1.

Table 3.1

	Mathematics	English
Mean	56·4	52·8
SD	16	12

To compare your relative performance with Moira's we need to first transform the two sets of marks to new ones having equal means and equal standard deviations. We can choose any values we like to be the common mean and common standard deviation of the two sets of transformed marks. A common mean of 0 and a common standard deviation of 1 are the easiest; that is, we convert the two sets of marks into standardized scores.

From (3) on the opposite page, the standardized score corresponding to M marks in Mathematics is given by

$$z_M = \frac{M - 56{\cdot}4}{16},$$

while the standardized score corresponding to E marks in English is given by

$$z_E = \frac{E - 52{\cdot}8}{12}.$$

It follows that your standardized score (M = 66) is

$$\frac{66 - 56{\cdot}4}{16} = 0{\cdot}6,$$

and Moira's standardized score (E = 66) is

$$\frac{66 - 52{\cdot}8}{12} = 1{\cdot}1.$$

Since Moira's standardized score is higher than yours, her relative performance in English is better than yours in Mathematics.

When a prize is to be awarded to the pupil who has shown the best overall performance in examinations involving several subjects, it is common practice to simply total the marks obtained by each pupil and to award the prize to the pupil with the highest total. This can be unfair when, as is often the case, the means and the standard deviations of the marks in the individual subjects are very different. A fairer procedure is to modify the marks in all the subjects so that the transformed marks have equal means and equal standard deviations and then total the pupils' modified marks. This is illustrated in the following example.

○ EXAMPLE 3.11

Four pupils, A, B, C, and D, in a class are in the running for a prize to be given for the best overall performance in Mathematics, Statistics, and Computing. The marks obtained by the four pupils in the three subjects and the means and the standard deviations of the marks of the entire class in the three subjects are shown in Table 3.2. Since all four pupils

Table 3.2

Subject	Pupil A	B	C	D	Class marks Mean	SD
Mathematics	60	50	60	70	68	12
Statistics	80	60	70	80	75	15
Computing	50	80	60	40	62	20
Total	190	190	190	190		

have the same total of marks in the three subjects, it would appear that their overall performances are equally good. To compare their relative performances, we shall first obtain the standardized scores of the marks in each subject. Using (3) from page 28 the standardized scores (z-values) are obtained as shown in Table 3.3a.

Table 3.3a

Subject	Marks	z-value
Mathematics	M	$(M-68)/12$
Statistics	S	$(S-75)/15$
Computing	C	$(C-62)/20$

Using these results the standardized scores corresponding to the marks obtained by the four pupils are shown in Table 3.3b.

Table 3.3b *Standardized Scores*

Subject	Pupil A	B	C	D
Mathematics	−0·67	−1·50	−0·67	+0·17
Statistics	+0·33	−1·00	−0·33	+0·33
Computing	−0·60	+0·90	−0·10	−1·10
Total	−0·94	−1·60	−1·10	−0·60

Ranking the totals from largest to smallest we see that the best relative overall performance was by D (−0.60), the second best by A (−0.94), the third best by C (−1.10), and the worst by B (−1.60). This analysis shows that the prize should be awarded to D. Since many find positive numbers easier to compare, we could have added any number to each of the totals for ordering the relative performances. For example, adding 10 to each total gives the respective values

$$9·06, 8·40, 8·90, 9·40,$$

from which it is clearer that D has the best overall performance. Alternatively, we could use the method illustrated in Example 3.10(b) to transform the marks in each subject so that the modified marks in each has, say, mean 50 and standard deviation 5.

10 A pupil had a mark of 83 in a Mathematics examination and a mark of 67 in a Physics examination. The mean and the standard deviation of the marks by all candidates in the Mathematics examination were 74 and 15, respectively, while the corresponding values in the Physics examination were 60 and 9, respectively. In which of the two subjects did this pupil show the better performance?

11 There are 5 pupils in a class being considered for first and second prizes for the best overall performances in examinations in Biology, Physics, and Chemistry. The marks obtained by these pupils and the class means and standard deviations in the three subjects are shown in Table 3.4. Determine which of the pupils should be awarded the first and the second prizes if the awards are to be based on
(a) totalling their marks over the three subjects,
(b) standardizing the marks so that the means and standard deviations are equal before totalling.

Table 3.4

Pupil	Biology	Physics	Chemistry
A	47	68	54
B	50	72	51
C	65	60	45
D	53	52	60
E	40	84	48
Mean	55	62	57
SD	5	10	7·5

1 Turn to any page of any novel. Count and record the number of words per line in the first ten lines on the page, ignoring any part-line at the end of a paragraph. Determine the mode (if it exists), the mean, and the median number of words per line. Which of your answers do you regard as the most appropriate single value representation of the number of words per line?

2 How many times would you expect to have to throw a fair die (having its six faces numbered 1, 2, 3, 4, 5, 6 respectively) in order to get a score of 6 for the first time?
Carry out ten times the experiment of throwing a fair die until you get a score of 6 and record the number of throws you made each time. Calculate the mean number of throws you made and comment on its value relative to the answer you gave to the question asked above.
Discuss the appropriateness of the mean as a single value representation of your data. Suggest an alternative quantity that might be more appropriate here as a single value representation (that is, a measure of location) and calculate a measure of dispersion that may be used in conjunction with the alternative measure of location.

3 Write down the amount of pocket money you get each week and amounts some of your friends get, aiming for a total of 10 or more amounts. Determine the median of the amounts and calculate the mean absolute deviation from the median.
 If you have read the section on box-and-whisker plots on p. 25 then construct a box-and-whiskers plot as a summary of the amounts.

Review Exercises
Chapter 3

☐ **LEVEL 1**

1 The percentage marks obtained by a pupil taking 8 subjects in an end-of-year examination were:

36, 49, 62, 74, 25, 39, 53, 62.

Determine the median and the mean of these marks. If the mean is chosen to be the measure of location for these marks, name and calculate the values of two measures of dispersion, either of which is appropriate for use in conjunction with the mean.

2 The annual salaries bill of a small business company is as follows:

Director	£24 000,
Manager	£20 000,
Salesman	£11 000
Chief operator	£10 000,
3 Operators at	£6 000 each,
2 Secretaries at	£5 000 each
1 Apprentice	£3 000.

(a) Find the mode, the median and the mean of these 10 salaries.
(b) Which of these would you regard as the fairest indicator of the 'average' salary?

3 The typing speeds, in words per minute (w.p.m.), of 10 typists working in a large office are:

37, 45, 41, 37, 39, 41, 41, 38, 43, 40.

(a) Determine the mode, the median, the mean, the mean absolute deviation from the mean, and the semi-interquartile range of these speeds.
(b) Name and give the values of one measure of location and one measure of dispersion which may be used to summarize the data.

4 (a) State which characteristic feature of a data set is measured by
(i) the mode, (ii) the median, (iii) the standard deviation, (iv) the mean absolute deviation from the median.
(b) Calculate their values for the following data:

4, 0, 9, 3, 4, 2, 9, 4, 1, 6, 1.

5 The times, in minutes, taken to complete a certain bus journey on ten different occasions were:

11, 9, 13, 11, 10, 12, 10, 28, 15, 11.

(a) Determine the values of three measures of location for these times. (b) Calculate the quartiles of the data and hence find the value of a measure of dispersion of the times. (c) State which of your measures of location should be used in conjunction with your measure of dispersion.

☐ LEVEL 2

6 The times, in seconds, gained (plus values) and lost (minus values) by 10 watches over a period of 24 hours were:

1·1, 4·2, −2·3, 3·4, 0·7, −4·1, 3·9, −3·0, 3·1, 1·0.

Calculate the mean and the standard deviation of these times.

7 The mean of a set of 12 numbers is 5 and the sum of the squares of the 12 numbers is 500. Calculate the standard deviation of the numbers.

8 A set of x-values has mean 25 and standard deviation 10. Write down the mean and the standard deviation of the set of values obtained on using the transformation (a) y=3x, (b) y=2x+1.

9 The times, in minutes, taken by a housewife to complete her shopping in eight visits to a supermarket were:

25, 14, 22, 24, 19, 26, 40, 16.

(a) Calculate the mean and the variance of these times.
(b) The housewife travelled to the supermarket on a bus which arrived there at 15.00 hours on each of the eight visits. (i) Find, to the nearest minute, the mean time at which the housewife left the supermarket. (ii) The housewife always aimed to catch the 15·50 bus for the return journey. Find, to the nearest minute, the mean and the standard deviation of the lengths of the times she spent waiting for the return bus.

10 A train is scheduled to arrive at a railway station at 17.35. A regular commuter on the train noted the following arrival times of the train on 11 days:

17.42, 17.38, 17.34, 17.44, 17.40, 17.32
17.33, 17.39, 17.41, 17.30, 17.34.

(a) For these arrival times find the values of (i) the median, (ii) the mean, (iii) the mean absolute deviation from the mean, (iv) the standard deviation, and (v) the semi-interquartile range.
(b) Name and give the values of one measure of location and one measure of dispersion which may be used together to summarize the train's arrival times.

11 (a) Using a transformation, or otherwise, calculate (i) the mean and (ii) the variance of the following data:

4·03, 4·07, 4·04, 4·09, 4·12, 4·10,
4·02, 4·08, 4·04, 4·11.

(b) Find the transformation that will convert the above data set into one having mean 400 and variance 11·4.

12 The four numbers a, a+2, 2a+1, and x, have mean a+4. Find the value of x. Given also that the variance of the four numbers is 14, find the possible values of a.

13 Table 3.5 shows the means and the standard deviations of the marks obtained by a class of pupils in three subjects A, B and C. A particular pupil's marks in A, B and C were 57, 62, and 50, respectively. Rank the pupil's performances relative to the whole class.

Table 3.5

Subject	A	B	C
Mean	53	68	60
Standard deviation	4	8	10

☐ APPENDIX

We will show that:

$$\Sigma(x-\bar{x})^2 = \Sigma x^2 - \frac{1}{n}(\Sigma x)^2.$$

Consider a data set of n observations $x_1, x_2, \ldots, x_n$ having mean $\bar{x} = (\Sigma x)/n$. The square of the deviation of x_1 from $\bar{x}$ is:

$$(x_1 - \bar{x})^2 = x_1^2 - 2\bar{x}x_1 + \bar{x}^2,$$

with similar expressions for $x_2, x_3, \ldots, x_n$. Summing over all the observations we have:

$$\Sigma(x-\bar{x})^2 = \Sigma x^2 - 2\bar{x}\Sigma x + n\bar{x}^2.$$

But $\bar{x} = \Sigma(x)/n$, so that:

$$\Sigma(x-\bar{x})^2 = \Sigma x^2 - \frac{2}{n}(\Sigma x)(\Sigma x) + n\left(\frac{1}{n}\Sigma x\right)^2$$

$$= \Sigma x^2 - \frac{2}{n}(\Sigma x)^2 + \frac{n}{n^2}(\Sigma x)^2$$

$$= \Sigma x^2 - \frac{1}{n}(\Sigma x)^2.$$

4

DISCRETE FREQUENCY DISTRIBUTIONS

"The average number of children per household on the housing estate where I live is 2.36."

In this chapter we consider the presentation and analysis of a moderately large or large data set on a discrete variable.

4.1 Frequency Distributions

○ EXAMPLE 4.1

Consider the data on the numbers of sweets in 36 tubes of Smarties as displayed in Table 1.1, p. 2. For a better appreciation of this set of data we first compile a tally chart as described in Section 2.1, p. 6. The tally chart for the numbers of sweets in the 36 tubes of Smarties is shown in Table 4.1. We can then display the data more neatly as shown in Table 4.2, in which we have now ordered the variable values (i.e. the number of sweets per tube) from the smallest to the largest. Such a table is referred to as a

Table 4.1 *Tally Chart for the Number of Sweets in 36 Tubes of Smarties*

No. of sweets	Tally	Frequency
37	⦀⦀ ⦀⦀ ⦀⦀	12
38	⦀⦀ ⦀⦀ ⦀⦀⦀	13
40	⦀⦀	2
39	⦀⦀ ⦀	6
36	⦀⦀⦀	3
	Total	36

frequency distribution table; in this example the table displays the **frequency distribution** of the number of sweets in 36 tubes of Smarties.

A frequency distribution may be represented diagrammatically in the form of a **frequency bar diagram** (sometimes referred to as a **vertical line diagram**). Such a diagram for the data in Table 4.2 is shown in Figure 4.1. In this diagram a vertical line is drawn above each distinct value of the variable; the height of the line represents the frequency of that value.

Table 4.2 *Frequency Distribution of the Number of Sweets per Tube in 36 Tubes of Smarties*

No. of sweets in tube	No. of tubes (frequency)
36	3
37	12
38	13
39	6
40	2
	Total 36

Figure 4.1 *Frequency Bar Diagram for Table 4.2*

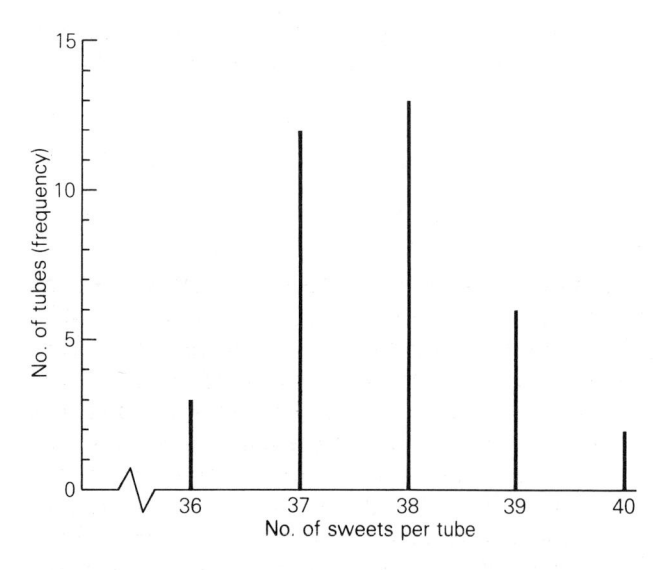

4.2 Summary Measures

The summary measures introduced in Chapter 3 are equally appropriate for discrete frequency distributions. Here we need only discuss the computational procedures for determining their values.

MEASURES OF LOCATION

THE MODE

Since the mode, when it exists, is the value in a data set that occurs most frequently, its existence and its value are easy to determine from a frequency distribution table. For the frequency distribution displayed in Table 4.2 it is immediately clear that there is a mode and that its value is 38, there being more tubes containing 38 sweets than any other number of sweets. If the numbers of sweets in our 36 tubes are representative of all tubes, then a purchased tube is more likely to contain 38 sweets than any other number. However, in our example we see that 12 of the tubes contained 37 sweets as compared with 13 that contained 38 sweets. Thus, even if our sample is fairly representative, a safer statement to make is that a purchased tube is most likely to contain 37 or 38 sweets.

In some instances the mode has particular significance as a single value representation of a data set. For example, the modal size of adult clothes or shoes would be very valuable information to a manufacturer wishing to concentrate on the mass production of just one size of item. As another example, the modal family size is a very relevant statistic when planning a housing estate.

● EXERCISE 4.1

Referring to Table 1.2, p. 3 compile a frequency distribution table and draw a frequency bar diagram for each of the following sets of data. Use a separate sheet for each set of data. Keep your diagrams for use in later exercises.

1 The shoe sizes of the 20 boys.
2 The shoe sizes of the 16 girls.
3 The number of boys per family in the 36 families; assume that there are no brothers in the class. (Why is this assumption necessary?)
4 The number of children per family in the 36 families. What assumptions do you need to make here?

● EXERCISE 4.2

Questions 1–4 Write down the mode (if it exists) for each of the distributions in Questions 1–4, Exercise 4.1.

THE MEAN

The mean of a data set is the value obtained on dividing the sum of all the observations in the data set by the number of observations in the data set (see

p. 19). From Table 4.2, the sum of all the 36 observed numbers of sweets in the tubes is given by:

$$(3 \times 36) + (12 \times 37) + (13 \times 38) + (6 \times 39) + (2 \times 40) = 1360,$$

since there were 3 tubes that contained 36 sweets, 12 that contained 37 sweets, and so on. Thus the mean number of sweets per tube is:

$$\bar{x} = \frac{1360}{36} = 37\tfrac{7}{9}.$$

In the general case, if x denotes an arbitrary one of the *distinct* values in a data set and f denotes the frequency of that value then the mean is given by the formula:

$$\bar{x} = \frac{1}{n}\Sigma f x,$$

the summation Σ being over all the distinct values and $n = \Sigma f$ being the total number of observations in the data set.

On a calculator having an accumulating memory key (usually labelled M+), the value of $\Sigma f x$ may be obtained by carrying out the series of operations shown in Figure 4.2. Check that the content of M+ is zero first.

Figure 4.2

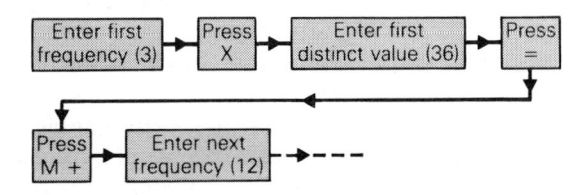

[Flowchart: Enter first frequency (3) → Press X → Enter first distinct value (36) → Press = → Press M+ → Enter next frequency (12) → ...]

The figures shown in brackets are those relating to the data in Table 4.2. Having completed these operations for the entire frequency distribution, pressing the Recall key (usually labelled RCL) will show the value of $\Sigma f x$ on the display.

As we found in Section 3.3, p. 25 the computation of $\bar{x}$ can be made easier by using a transformation (a working origin and scale) to produce simpler data values. For the data of Table 4.2 one obvious choice of transformation is to take $y = x - 36$. The corresponding frequency distribution of the y-values is then as displayed in Table 4.3 and its mean is given by:

$$\bar{y} = \frac{(3 \times 0) + (12 \times 1) + (13 \times 2) + (6 \times 3) + (2 \times 4)}{36} = \frac{64}{36} = 1\tfrac{7}{9}.$$

Table 4.3 *Frequency Distribution of* $y = x - 36$

y-value	0	1	2	3	4
Frequency	3	12	13	6	2

Since the transformation used only consisted of subtracting 36 from each x-value, the mean of the x-values is given by:

$$\bar{x} = \bar{y} + 36 = 37\tfrac{7}{9},$$

as calculated earlier.

For the general transformation:

$$y = (x - c)/k,$$

the value of $\bar{x}$ may be deduced from that of $\bar{y}$ using the formula:

$$\bar{x} = k\bar{y} + c,$$

obtained by reversing the order and inverting the operations applied to x in order to get y. (See (3.4) p. 27.)

Questions **5–8** *Find the mean of each of the distributions in Questions 1–4, Exercise 4.1.*

☐ THE MEDIAN

In a previous section (*The Median*, p. 19) we defined the median of a data set to be the value of the middle observation when the observations are ranked from smallest to largest, the convention being to take the average of the two middle observations when the number of observations in the data set is even.

Let us now determine the median of the frequency distribution in Table 4.2. Since there are 36 observations, the median is the average of the 18th and the 19th observations. Accumulating the frequencies in Table 4.2 from left to right we see that $3 + 12 = 15$ observations are 37 or less, and 13 observations are equal to 38. It follows that both the 18th and the 19th observations have the value 38, so that the median number of sweets per tube is:

$$m = 38.$$

It may happen in some examples that the two middle values in a data set of an even number of observations are different, in which case we average them to obtain the median. This is illustrated in the following example.

○ EXAMPLE 4.2

Table 4.4 shows the frequency distribution of the number of children per family in 30 families. Find the median number of children per family.

In this data set $n = 30$, so that the median is the average of the 15th and the 16th observations. From

Table 4.4 *Frequency Distribution of the Number of Children in 30 Families*

No. of children	0	1	2	3	4	Total
No. of families	6	9	11	3	1	30

Table 4.4, $6+9=15$ families have up to 1 child, 11 families have 2 children. Thus the 15th observation has the value 1 while the 16th has the value 2. Hence the median number of children per family is given by:

$$m = \tfrac{1}{2}(1+2) = 1\tfrac{1}{2}.$$

Questions **9–12** *Determine the median of each of the distributions in Questions 1–4, Exercise 4.1.*

☐☐ MEASURES OF DISPERSION

☐ WHEN THE MEAN IS THE MEASURE OF LOCATION

An appropriate measure of dispersion to use in conjunction with the mean as the measure of location is either MAD($\bar{x}$), the mean absolute deviation from the mean $\bar{x}$, or SD, the standard deviation, as defined in Chapter 3.

☐ THE MEAN ABSOLUTE DEVIATION FROM THE MEAN

For the data of Table 4.2 we have already established that the mean is $\bar{x} = 37\tfrac{7}{9}$. To evaluate MAD($\bar{x}$) we first need to calculate the sum of the absolute deviations of the observations from $\bar{x}$, and then divide this sum by 36. Evaluating the absolute deviation from $\bar{x}$ for each of the distinct values in Table 4.2 we find that the frequency distribution of the absolute deviations is as shown in Table 4.5.

For example, there are 3 observations for which the absolute deviation is $1\tfrac{7}{9}$, 12 for which it is $\tfrac{7}{9}$, and so on. Thus the mean absolute deviation is given by:

MAD($\bar{x}$) =

$$\frac{(3 \times 1\tfrac{7}{9}) + (12 \times \tfrac{7}{9}) + (13 \times \tfrac{2}{9}) + (6 \times 1\tfrac{2}{9}) + (2 \times 2\tfrac{2}{9})}{36}$$

On multiplying both the numerator and the denominator by 9 we have:

MAD($\bar{x}$) =

$$\frac{(3 \times 16) + (12 \times 7) + (13 \times 2) + (6 \times 11) + (2 \times 20)}{(9 \times 36)}$$

$$= 0.81 \quad \text{(to 2 d.p.).}$$

Table 4.5 *Frequency Distribution of the Absolute Deviations from the Mean for the Data in Table 4.2*

Absolute deviation from $\bar{x}$	Frequency
$1\tfrac{7}{9}$	3
$\tfrac{7}{9}$	12
$\tfrac{2}{9}$	13
$1\tfrac{2}{9}$	6
$2\tfrac{2}{9}$	2
Total	36

Questions **13–16** *Calculate the mean absolute deviation from the mean of each of the distributions in Questions 1–4, Exercise 4.1.*

☐ THE STANDARD DEVIATION

You may recall (from p. 23) that the standard deviation of a data set of n observations is usually more conveniently calculated using the expression:

$$SD = \frac{1}{n}\sqrt{(n\sum x^2 - (\sum x)^2)}$$

For a frequency distribution in which the value x occurs with frequency f, this becomes:

$$SD = \frac{1}{n}\sqrt{(n\sum fx^2 - (\sum fx)^2)} \qquad (4.1)$$

where $n = \Sigma f$.

Consider the frequency distribution of the number of sweets in 36 tubes of Smarties, given in Table 4.2 which, for our convenience, is reproduced below.

No. of sweets (x)	36	37	38	39	40	Total
No. of tubes (f)	3	12	13	6	2	36

For these data $n = 36$,

$$\Sigma fx = (3 \times 36) + (12 \times 37) + (13 \times 38) + (6 \times 39) + (2 \times 40) = 1360,$$

and

$$\Sigma fx^2 = (3 \times 36^2) + (12 \times 37^2) + (13 \times 38^2) + (6 \times 39^2) + (2 \times 40^2) = 51\,414.$$

Using Expression (4.1) we then have:

$$SD = \frac{\sqrt{(36 \times 51\,414 - 1360^2)}}{36}$$

$$= \frac{\sqrt{1304}}{36} = 1.00 \quad \text{(to 2 d.p.).}$$

Figure 4.3

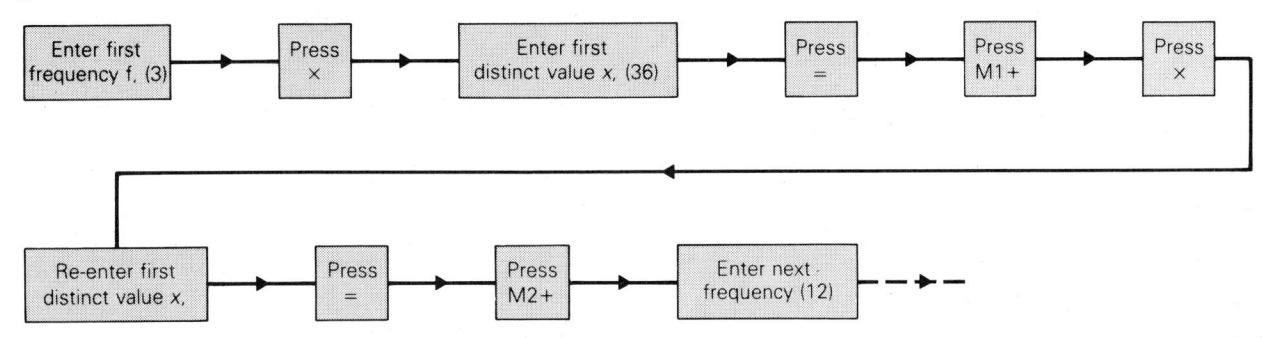

[Flowchart: Enter first frequency f, (3) → Press × → Enter first distinct value x, (36) → Press = → Press M1+ → Press × → Re-enter first distinct value x, → Press = → Press M2+ → Enter next frequency (12) → ...]

Some of the more specialized calculators are preprogrammed to give the values of Σfx and Σfx^2 from just one entry of the pairs of values (f, x). This facility is particularly useful when using Expression (4.1) to calculate the standard deviation of a frequency distribution. You will need to refer to your manual to check if this facility is available on your calculator. On a less specialized calculator it is possible to evaluate Σfx and Σfx^2 provided the calculator has two accumulating memories, usually labelled as M1+ and M2+, but check your manual. In this case the procedure to use is as shown schematically in Figure 4.3. First check that both M1+ and M2+ currently contain zero.

The figures in brackets are the values relating to the data of Table 4.2. Proceeding in this way for all (f, x) the ultimate content of M1+ will be the value of Σfx, while that of M2+ will be the value of Σfx^2. Pressing the corresponding recall keys RCL1 and RCL2 will show these values on the display.

If your calculator has only one accumulating memory then Σfx and Σfx^2 will have to be calculated separately. The operations for calculating Σfx were given in Figure 4.2; those for calculating Σfx^2 are shown in Figure 4.4. Pressing the recall key will then show the value of Σfx^2 on the display.

Figure 4.4

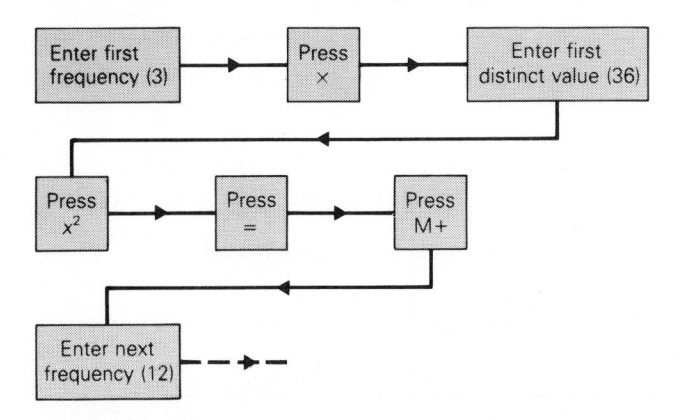

[Flowchart: Enter first frequency (3) → Press × → Enter first distinct value (36) → Press x^2 → Press = → Press M+ → Enter next frequency (12) → ...]

The arithmetic involved in calculating the standard deviation of the data in Table 4.2. may have been simplified on, for example, subtracting 36 from each x-value; that is, by transforming to $y = x - 36$. As was shown on p. 26 under this transformation $SD(x) = SD(y)$.

For the more general transformation:

$$y = (x - c)/k,$$

the mean and the standard deviation of the frequency distribution of x may be deduced from those of the frequency distribution of y using the formulae:

$$\bar{x} = k\bar{y} + c,$$

and:

$$SD(x) = kSD(y).$$

Questions **17–20** *Calculate the standard deviation of each of the distributions in Questions 1–4, Exercise 4.1.*

☐ WHEN THE MEDIAN IS THE MEASURE OF LOCATION

Appropriate measures of dispersion to use in conjunction with the median as the measure of location are MAD(m), the mean absolute deviation from the median m, and SIQR, the semi-interquartile range.

☐ THE MEAN ABSOLUTE DEVIATION FROM THE MEDIAN

We showed earlier that the median of the data in Table 4.2 is $m = 38$. You may recall from p. 22 that MAD(m), the mean absolute deviation from the median, is the average of the absolute deviations of the observations from their median m. From Table 4.2 the frequency distribution of the absolute deviations from $m = 38$ are as shown in Table 4.6. The sum of these absolute deviations is:

$$(3 \times 2)+(12 \times 1)+(13 \times 0)+(6 \times 1)+(2 \times 2)=28,$$

so that the mean absolute deviation from $m=38$ is given by:

$$\text{MAD}(m)=\frac{28}{36}=0\cdot78 \quad \text{(to 2 d.p.).}$$

Table 4.6 *Frequency Distribution of the Absolute Deviations from* $m=38$ *for the Data in Table 4.2*

Variable value (x)	Absolute deviation from 38	Frequency
36	2	3
37	1	12
38	0	13
39	1	6
40	2	2
Total		36

Questions **21–24** *Calculate the mean absolute deviation from the median for each of the distributions in Questions 1–4, Exercise 4.1.*

☐ THE SEMI-INTERQUARTILE RANGE

You may recall from p. 24 that the lower quartile (LQ) and the upper quartile (UQ) of a data set of n observations ordered from the smallest to the largest were defined as follows:

(a) If $n/4=k$ plus a remainder (1, 2, or 3) then:
    LQ is the value of the $(k+1)$th observation,
    UQ is the value of the $(n-k)$th observation.
(b) If $n/4=k$ with no remainder then:
    LQ is the average of the values of the kth and the $(k+1)$th observations,
    UQ is the average of the values of the $(n-k)$th and the $(n-k+1)$th observations.

For the data of Table 4.2, $n/4=9$, so that $k=9$ and $n-k=27$. Applying (b) above and referring to Table 4.2 or Table 4.6, we find that the quartiles are as follows. LQ is the average of the values of the 9th and 10th observations:

$$\text{LQ}=\tfrac{1}{2}(37+37)=37.$$

UQ is the average of the values of the 27th and 28th observations:

$$\text{UQ}=\tfrac{1}{2}(38+38)=38.$$

It follows that the semi-interquartile range (SIQR) of the distribution is given by:

$$\text{SIQR}=\tfrac{1}{2}(38-37)=\tfrac{1}{2}.$$

The distribution of the number of sweets in the 36 tubes of Smarties could now be summarized in the form of a box-and-whisker plot similar to Figure 3.6, p. 25.

As an illustration of (a) above consider the data of Table 4.4, the frequency distribution of the number of children per family in 30 families. For convenience the table is reproduced here:

No. of children	0	1	2	3	4	Total
No. of families	6	9	11	3	1	30

In this data set, $n/4=7+\tfrac{1}{2}$, so that $k=7$ and $n-k=23$. Referring to the table above we find that the quartiles are as follows: LQ is the value of the 8th observation:

$$\text{LQ}=1.$$

UQ is the value of the 23rd observation:

$$\text{UQ}=2.$$

so the semi-interquartile range of the distribution is:

$$\text{SIQR}=\tfrac{1}{2}(2-1)=\tfrac{1}{2}.$$

Questions **25–28** *Calculate the semi-interquartile range of each of the distributions in Questions 1–4, Exercise 4.1.*

29 *Table 4.7, extracted from the 1981 Census Report for the county of Dyfed (Wales), shows the frequency distribution of the number of*

Table 4.7

No. of persons	No. of households (thousands)
1	24
2	37
3	21
4	21
5	9
6	3
7+	1
Total	116

persons usually resident in households that are permanent buildings.
(a) Determine the median and the semi-interquartile range of this distribution.
(b) *Summarize the distribution in the form of a box-and-whisker diagram.
(c) Why is it not possible to find the mean and the standard deviation exactly from the distribution given?

No. of sweets in tube (x)	No. of tubes (frequency, f)
36	3
37	12
38	13
39	6
40	2
	Total 36

4.3 A GRAPHICAL METHOD FOR DETERMINING QUARTILES

It was mentioned in the section on *The Semi-Interquartile Range* on p. 24 that is is not generally possible to find three observations in a data set which will divide the data set into four exactly equal quarters, and that for this reason various definitions have been proposed for finding three values which attain this aim approximately. One definition was given in the previous section. Whichever definition is used, the three values obtained are referred to as the quartiles of the data set, even though the various definitions do not always lead to the same values.

Here we shall describe a graphical method for determining the quartiles of a frequency distribution. This gives values that are identical with those obtained from the definitions in the previous section. This method also has the advantage that it carries over very nicely to data on a continuous variable, as will be shown in the next chapter.

The first step in the graphical method is to construct a **cumulative frequency distribution**. For each distinct value of the variable, the total number of observations that are less than or equal to it is called the **cumulative frequency** of that value. The cumulative frequencies are obtained from a frequency distribution simply by successively adding the frequencies from left to right in the order in which they appear in the frequency distribution table.

○ *EXAMPLE 4.3*

Consider the data of Table 4.2, reproduced again here:

Here:
  the cumulative frequency of 36 is 3,
  the cumulative frequency of 37 is $3+12=15$,
  the cumulative frequency of 38 is $15+13=28$,
  the cumulative frequency of 39 is $28+6=34$,
  the cumulative frequency of 40 is $34+2=36$
which is the total frequency.

The cumulative frequency distribution table is shown in Table 4.8, where we use the abbreviation CF for cumulative frequency.

Table 4.8 *Cumulative Frequency Distribution Table for the Number of Sweets in 36 Tubes of Smarties*

No. of sweets (x)	Cumulative frequency (CF)
36	3
37	15
38	28
39	34
40	36

We now consider a graphical representation of these data. Some care is needed. In particular, observe that for a value of x from 36 up to but not including 37 the cumulative frequency remains constant at the value 3, and that when $x=37$ the cumulative frequency jumps up to the value 15. Similarly, for any value of x from 37 up to but not including 38, the cumulative frequency remains constant at the value 15, and when $x=38$ the cumulative frequency jumps up to the value 28; and so on for the other values of x up to and including 40. The graph of the cumulative frequency distribution will therefore consist of steps (or jumps), the rises in these steps occurring at the observed values of the variable. The resulting graph for Table 4.8 is displayed in Figure 4.5. and is referred to as a cumulative frequency **step polygon**.

[39

Figure 4.5 *Cumulative Frequency Step Polygon for Table 4.8*

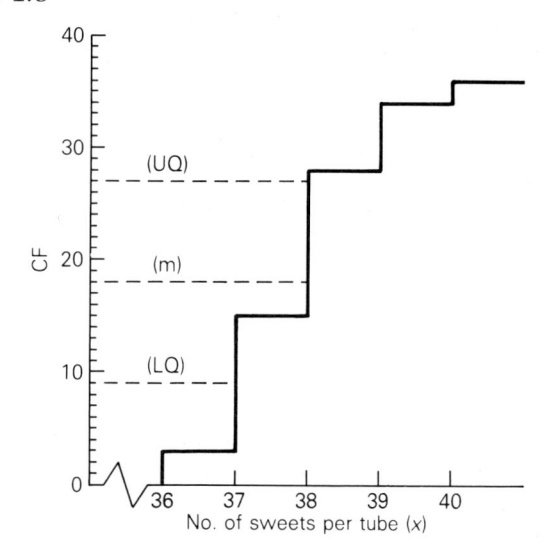

To determine the median of the distribution from this graph we read off the variable value for which $CF = \frac{1}{2}n$. In our example $n = 36$, so that the median is the value of x for which $CF = \frac{1}{2} \times 36 = 18$. We have drawn a broken line at this value in Figure 4.5, and we find that the median is:

$$m = 38,$$

as obtained previously. To determine the lower quartile LQ we read off the variable value for which $CF = \frac{1}{4}n = \frac{1}{4} \times 36 = 9$ in our example. From the graph we see that:

$$LQ = 37,$$

again agreeing with the result we obtained previously. Finally, to find the upper quartile UQ we read off the variable value for which $CF = \frac{3}{4}n = \frac{3}{4} \times 36 = 27$ in our example. From the graph we find that:

$$UQ = 38,$$

as obtained previously.

When using this graphical method it may happen that the horizontal line drawn from the CF value of interest (corresponding to the median or either of the lower and upper quartiles) actually runs directly into a horizontal line on the graph. In such a case, the corresponding x-value is taken to be the average of the two values of x at the ends of the horizontal line. This is illustrated in the next example.

○ **EXAMPLE 4.4**

Consider the data in Table 4.4, the frequency distribution of the numbers of children in 30 families. For ease of reference the table is reproduced here. The corresponding cumulative frequency dis-

No. of children (x)	0	1	2	3	4
No. of families (f)	6	9	11	3	1

tribution table is displayed in Table 4.9 and graphed in Figure 4.6.

Since $n = 30$ in this example, the median is the variable value for which $CF = \frac{1}{2} \times 30 = 15$. We note

Table 4.9 *Cumulative Frequency Distribution Table for the Numbers of Children in 30 Families*

No. of children (x)	Cumulative frequency (CF)
0	6
1	15
2	26
3	29
4	30

Figure 4.6 *Cumulative Frequency Graph for Table 4.9 (Step Polygon)*

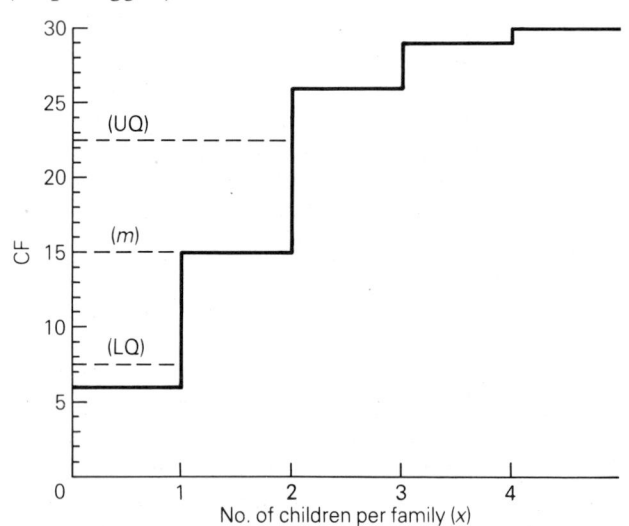

from Figure 4.6 that the horizontal line for which $CF = 15$ runs directly into the horizontal line on the graph above the interval from $x = 1$ to $x = 2$, so that the median is $m = \frac{1}{2}(1 + 2) = \frac{3}{2}$, agreeing with the value calculated earlier. Since $\frac{1}{4} \times 30 = 7\frac{1}{2}$ and $\frac{3}{4} \times 30 = 22\frac{1}{2}$, it follows from the graph in Figure 4.6 that the quartiles are $LQ = 1$ and $UQ = 2$. These may be verified to be the values of the 8th and the 23rd observations, respectively, as given by the rules we used earlier.

As an alternative to plotting the cumulative frequencies against the variable values we could plot the **cumulative relative frequencies**. You may recall from Section 2.4, p. 12 that the relative frequency of a variable value in a data set is simply its frequency divided by the number of observations in the data set (the total frequency). Adding the relative frequencies in the same manner as we added the frequencies will give the cumulative relative frequencies of the variable values.

An advantage of using the cumulative relative frequencies, is that in a graph the vertical axis used for the cumulative relative frequencies will always be scaled from 0 to 1 irrespective of the number of observations in the data set. The median is then the variable value for which the cumulative relative frequency (CRF) is 0.5, and the quartiles are the variable values for which CRF = 0.25 and CRF = 0.75, respectively. Another advantage arises if one wishes to give a graphical comparison of two frequency distributions when the numbers of observations in the data sets are different.

● EXERCISE 4.3

Questions *1–5* For each of the data sets in Exercise 4.1 and Question 29, Exercise 4.2 p. 38, use the graphical method described above to determine the median and the quartiles. Summarize each data set in the form of a box-and-whisker plot.

4.4 DECILES AND PERCENTILES

The five quantities – smallest observation, largest observation, median, and the quartiles – whose values are shown in a box-and-whisker plot, provide a very informative summary of a data set, particularly when the number of distinct values in the data set is not very large. When the number of distinct values is large it may well be desirable to have a more informative summary than is provided by the five values given in a box-and-whisker plot. A convenient way of introducing additional quantities is to divide the data set into more than four equal parts. Dividing a data set into 10 equal parts will produce nine values which are referred to as the **deciles**, while dividing the data set into 100 equal parts will produce 99 values which are called the **percentiles**.

We shall denote the deciles of a data set by $D_1, D_2, \ldots, D_9$, and the percentiles by $P_1, P_2, \ldots, P_{99}$, in each case assuming them to be in increasing order of magnitude. Observe that $D_5 = P_{50} = m$ (the median), that $D_1 = P_{10}$, $D_2 = P_{20}$, etc., and that $P_{25} = $ LQ, the lower quartile, and $P_{75} = $ UQ, the upper quartile.

For a data set of n observations, the percentile P_r, where r is an integer from 1 to 99, inclusive, is defined as follows:

(a) If $rn/100 = k + e$, where k is a whole number and e is fractional, then P_r is the value of the $(k+1)$th observation.
(b) If $rn/100 = k$, where k is a whole number, then P_r is the average of the values of the kth and $(k+1)$th observations.

(Observe that these definitions agree with those given on p. 24 for the lower and upper quartiles on setting $r = 25$ and $r = 75$, respectively.)

As an alternative to using these rules we can also determine percentiles graphically as illustrated in the following example.

○ EXAMPLE 4.5

The frequency distribution of the number of defective items produced per day at a factory over a period of 85 days is shown in Table 4.10. Determine the values of the 60th and the 90th percentiles of the distribution.

Table 4.10 *Frequency Distribution of the Number of Defective Items Produced per Day over a Period of 85 Days*

No. of defectives	No. of days
0	22
1	18
2	10
3	7
4	6
5	6
6	5
7	4
8	3
9	1
10	1
11	1
12	1
Total	85

We first construct the cumulative frequency distribution shown in Table 4.11, where x denotes the number of defective items produced per day and CF denotes the cumulative frequency of the associated x-value (that is, the number of days for which the total number of defective items is less than or equal to the x-value). The graph of this cumulative frequency distribution is displayed in Figure 4.7. (Alternatively, as mentioned at the end of Section 4.3, we could have plotted cumulative relative frequencies instead of cumulative frequencies.)

Table 4.11 *Cumulative Frequency Distribution for Table 4.10*

x	CF
0	22
1	40
2	50
3	57
4	63
5	69
6	74
7	78
8	81
9	82
10	83
11	84
12	85

Figure 4.7 *Cumulative Frequency Graph for Table 4.10 (Step Polygon)*

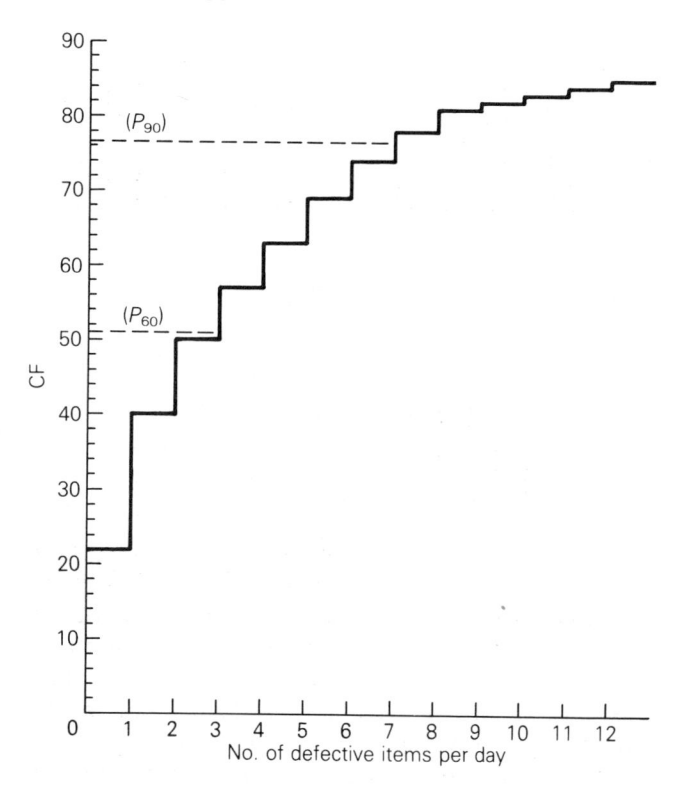

The 60th percentile is the number of defective items for which CF is 60% of n so CF = 0·6 × 85 = 51, which we have indicated as P_{60} on the graph. Reading from the graph we find that:

$$P_{60} = D_6 = 3.$$

The 90th percentile is the number of defective items for which CF is 90% of n, so CF = 0·9 × 85 = 76·5, which is indicated on the graph as P_{90}. Reading from the graph we have:

$$P_{90} = D_9 = 7.$$

The difference between two percentiles corresponding to percentages that add up to 100 is called an **interpercentile range** and is often used as a measure of dispersion. The interquartile range (UQ–LQ), which is equal to (P_{75}–P_{25}), is one possibility that we have already mentioned. Other interpercentile ranges that are sometimes used include (P_{90}–P_{10}), (P_{70}–P_{30}), and (P_{60}–P_{40}). Let us evaluate (P_{60}–P_{40}), the 40–60 interpercentile range for the data in Table 4.10, whose cumulative frequency graph is given in Figure 4.10. Now P_{40} is the variable value for which CF is 40% of 85. So CF = 0·4 × 85 = 34, and from the graph we find that this value is 1. Since P_{60} = 3 (shown earlier), it follows that the 40–60 interpercentile range is (P_{60}–P_{40}) = 3 – 1 = 2.

● EXERCISE 4.4

1 *Table 4.12 gives the frequency distribution of the number of particles emitted from a radioactive source in 100 successive periods each of duration one minute. Determine (a) the median number of emissions per minute, (b) the 20–80 interpercentile range of the distribution.*

Table 4.12

No. of emissions	No. of periods
12	2
13	4
14	9
15	14
16	18
17	20
18	13
19	10
20	6
21	4

2 *Table 4.13 shows the frequency distribution of the number of units used per telephone call by a person during a quarter when the total number of calls made by that person was 110.*
(a) Determine the values of the deciles D_3, D_4, D_5 and D_6.
(b) Suggest and evaluate a measure of dispersion

of the distribution based on some or all of these values.

Table 4.13

No. of units used	No. of calls
3	18
4	20
5	22
6	11
7	13
8	9
9	5
10	6
11	5
12	1

3 *Determine the 40th and the 60th percentiles of the frequency distribution of Question 29, Exercise 4.2, p. 38.*

4.5 Distribution Shapes

Frequency distributions, as portrayed by their bar diagrams, arise in assorted shapes, some of which we have already met in examples and exercises.

One of the most common shapes that arises in practice is that in which, as the variable value increases, the associated frequencies start to increase to a maximum (at the mode) and then decrease, as in Examples 4.1 and 4.2. A distribution which has a mode is said to be **unimodal** and if its bar diagram is such that the part to one side of the mode is the mirror image of the part to the other side of the mode, the distribution is said to be **symmetrical**; an example of such a distribution is given in Figure 4.8a. In Figure 4.8b we illustrate a distribution which is symmetrical but not unimodal. A truly symmetrical frequency distribution occurs rarely; Figure 4.1 p. 34 shows a frequency distribution which is almost symmetrical.

A distribution which is not symmetrical is said to be **skew**. If its bar diagram is such that the 'tail' to the right of the mode is longer than that to the left of the mode, the distribution is said to be **positively skew**; an example is shown in Figure 4.9a. If the bar diagram is such that the 'tail' to the left of the mode is longer than that to the right of the mode, the distribution is said to be **negatively skew**, an example of which is shown in Figure 4.9b. The J-shaped and reverse J-shaped distributions illustrated in Figures 4.9c and d are the most extreme skew distributions. Other shapes may arise in practice but we shall not detail all the possibilities here.

Figure 4.9 *Examples of Skew Distributions: (a) Positively skew, (b) Negatively skew, (c) J-shaped, (d) Reverse J-shaped*

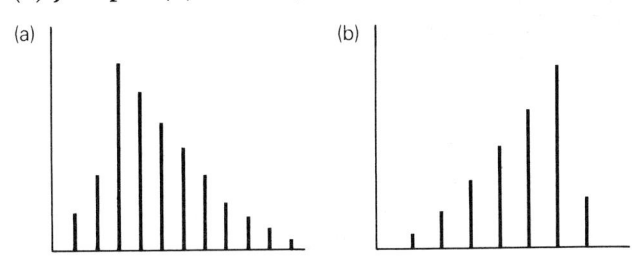

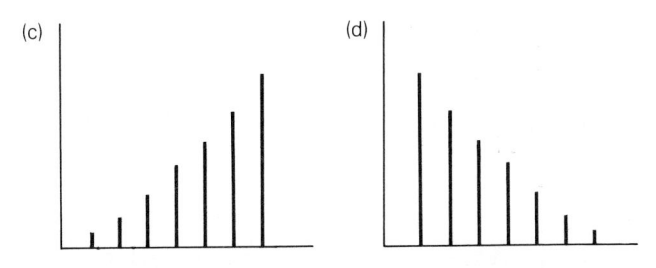

When a distribution is symmetrical its mean and median will be equal. If in addition the distribution is unimodal, then the mode, the median and the mean will all be equal. For a skew distribution the mode and the median may or may not be equal; in either case the mean will have a value to the side of the mode in the direction of the longer 'tail'.

For a skew distribution whose mode and median are unequal, the median will lie between the mean and the mode; the magnitudes of the mean, median and mode will have the same order (or reverse order) as the words have in a dictionary. To see this, consider a positively skew distribution like Figure 4.9a. The relatively large values of the variable will have the effect of inflating the mean to a value to the right of the mode but will have little or no effect on

Figure 4.8 *Examples of Symmetrical Distributions: (a) Unimodal and Symmetrical, (b) Symmetrical But Not Unimodal*

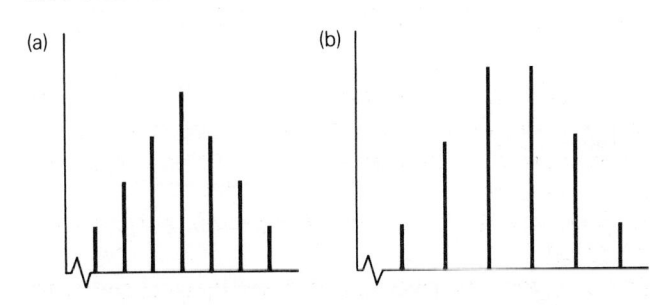

[43

Doleful data

Think of different ways of classifying employment data, e.g. by type of worker, by month when unemployment benefit is first collected, by length of time unemployed, etc. Which types of classification fit the methods which you have met in chapters 2, 3 and 4? Supposing you wanted to work out what the median length of unemployment was in a particular district, how would you collect the data and organise the frequency table and diagrams? Can you see how the way in which the data are collected might influence whether "length of unemployment" is a continuous or a discrete variable? What happens if you ask people how long it is since they last worked as opposed to asking them how many weeks' benefit they have received?

the value of the median. Thus if the mode and median are different, the values of the three measures of location will be such that the mode is least, the mean is greatest and the median lies between the mode and the mean. Likewise, for a negatively skew distribution whose mode and median are different, the values of the three measures of location in increasing order of magnitude will be mean, median and mode.

One consequence of all this is that for a very skew distribution the mean may be inappropriate as a single value representation of the distribution and one should use the mode or the median.

Furthermore, the relatively large (or small) values in a skew distribution will also inflate both the mean absolute deviations and the standard deviation, so that none of these is then appropriate as a measure of dispersion for a skew distribution. If follows that when a distribution is very skew the appropriate choices of measures of location and dispersion are the median and semi-interquartile range. This would be particularly so for J-shaped or reverse J-shaped distributions like those shown in Figures 4.9c and d.

● **EXERCISE 4.5**

Questions *1–4* For each of the distributions in Questions 1–4, Exercise 4.1, state whether it is (a) symmetrical, (b) almost symmetrical, (c) positively skew, or (d) negatively skew.

5 A data set on a discrete variable gave the frequency distribution shown in Table 4.14.

Table 4.14

Variable value	10 11 13 14 20 40
Frequency	12 16 8 4 4 4

(a) Draw a frequency bar diagram.
(b) State the shape of the distribution.
(c) Determine the mode, the median and the mean. How do their relative values reflect the shape of the distribution?
(d) Determine the values of a measure of location and a measure of dispersion that are appropriate for this distribution.

6 A count was made of the number of matches contained in each of 53 boxes of matches. The results obtained are given in Table 4.15.
(a) Draw a frequency bar diagram to exhibit the distribution.
(b) State whether the distribution is almost symmetrical or very skew.
(c) Determine the mode, the median and the mean. In what sense do your values support your answer to (b)?

Table 4.15

No. of matches	43 44 45 46 47 48
No. of boxes	3 8 16 17 8 1

*Projects

1 Place four coins in your hand. Cup your hands together, shake the coins and then allow them to fall onto a flat surface. Record the number of coins that show heads. Carry out this experiment a total of 50 times.
(a) Present the results you obtain in the form of a frequency distribution and display the distribution in a diagram.
(b) Calculate the mean number of heads that you obtained per throw. What would you have expected this value to be? Comment on any discrepancy.
(c)* Summarize your data in a box-and-whisker plot.

2 Repeat Project 1 but now use four dice and record the number of 6s you obtain in each throw.

3 Try to obtain a listing of the population sizes of at least 50 towns and cities. (One useful source for such data is the handbook of the Automobile Association (AA) which gives the population sizes of towns in England, Wales and Scotland.) Record the first and the third digits in each population figure, the first digit being one of 1, 2, ..., 9 and the third being one of 0, 1, 2, ..., 9.
(a) Present your results for the first digit and for the third digit as frequency distributions. In each case, discuss whether all the possible digits seem to occur about equally frequently.
(b)* Summarize the two distributions as box-and-whisker plots and comment on any differences between them.

(A theoretical study of the distribution of the first digit in a collection of numbers obtained by counting (as population sizes are) has shown that the proportion of the numbers whose first digit is n or less is approximately $\log_{10}(n+1)$. For example, the proportion of the numbers whose first digit is from 1 to 4 (inclusive) is approximately $\log_{10} 5 = 0.7$. Compare your proportion with this theoretical one.)

4 For this project you will need one copy of each of two novels, one popular paperback and another by a reputable author. Turn to any page in each of the novels and count the numbers of letters in the first 100 words. Display the two data sets as frequency distributions. For each data set calculate the mean and the standard deviation of the number of letters per word. Comment on any differences.

5 Married couples belonging to a certain tribe are forbidden to have any more children after the birth of their first son. Assuming that each child born is equally likely to be a boy or a girl, the sex of an unborn child may be likened to the outcome of a toss of a fair coin, the occurrence of a head corresponding to a girl and the occurrence of a tail corresponding to a boy.

Toss a coin until you get a tail and record the number of tosses you make; this number may be interpreted as the number of children a married couple will have. Carry out this experiment of tossing a coin until a tail is obtained a total of 50 times (say), each time recording the number of tosses made. You will thus have generated the numbers of children of 50 married couples.
(a) Present your results in the form of a frequency distribution of the number of children per married couple.
(b) Calculate the ratio of the total number of girls to the total number of boys in all 50 families.
(c) Calculate the proportions of the 50 families that have (i) exactly 1 daughter, (ii) exactly 2 daughters, (iii) exactly 3 daughters, (iv) 4 or more daughters.

(The answers you obtain will be estimates of the corresponding quantities for all the families in the tribe provided that multiple births are not possible and that neither parent dies before a son is born.)

Review Exercises Chapter 4

☐ LEVEL 1

1 Figure 4.10 shows the frequency distribution of the number of children per family living on a housing estate.

Figure 4.10

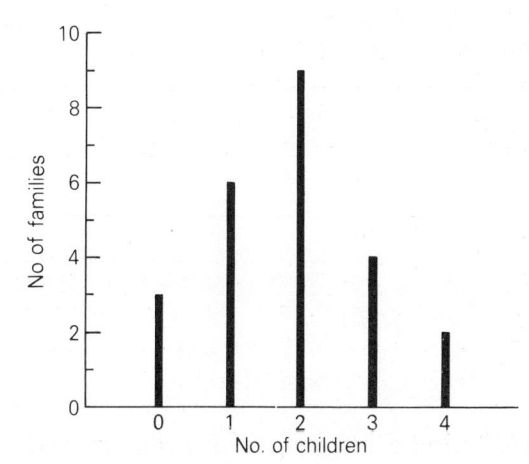

(a) Determine the number of families living on the estate.
(b) Write down the modal number of children per family.
(c) Determine the median number of children per family.
(d) Calculate the mean number of children per family.
(e) Calculate the mean absolute deviation from the median.
(f) From the answers you have given above choose two which may serve as values of a measure of location and associated measure of dispersion of the distribution.

2 Last season a particular football team played 24 league matches. The numbers of goals the team scored in these matches were as follows:

0, 1, 2, 1, 3, 0, 1, 4, 1, 1, 2, 1,
2, 1, 3, 0, 1, 2, 0, 3, 1, 1, 1, 4.

(a) Compile a table of the frequency distribution of the number of goals scored per match.
(b) Calculate the mean and, to two decimal places, the standard deviation of the numbers of goals scored per match.

3 The first-round scores of 40 golfers in a competition were as follows:

70, 68, 68, 67, 66, 68, 68, 66, 66, 67,
65, 68, 66, 70, 65, 69, 67, 68, 68, 66,
67, 70, 66, 68, 68, 69, 70, 67, 67, 68,
68, 69, 69, 69, 67, 68, 67, 69, 67, 69.

(a) Display the data as a frequency distribution.
(b) Present the distribution diagrammatically.
(c) Determine the median and the semi-interquartile range of the scores.
(d) Calculate another quantity which may be used in conjunction with the median as a measure of the dispersion of the scores.

4 Table 4.16 shows the distribution of the numbers of examination passes obtained by the 30 pupils in a class.

Table 4.16

No. of passes	3	4	5	6	7	8
No. of boys	1	2	4	2	1	2
No. of girls	3	2	2	6	4	1

(a) Show that the mean absolute deviation from the median has the same value for both the boys' and the girls' results.
(b) Calculate the mean and the standard deviation of the number of passes per pupil for the entire class.

5 The number of telephone calls received per minute at an exchange over a period of one hour had the distribution shown in Table 4.17.
Find graphically, or otherwise, the median and the semi-interquartile range of the distribution.

Table 4.17

No of calls	0	1	2	3	4
No of minute periods	5	17	23	14	1

6 Table 4.18 shows the frequency distribution of the numbers of defective items found in 100 batches of manufactured items.

Table 4.18

No. of defective items	No. of batches
0	23
1	28
2	16
3	14
4	11
5	5
6	2
7	1

(a) Write down the modal number of defective items per batch.
(b) Calculate the mean and the mean absolute deviation from the mean.

7 Inspection of a typed document consisting of 50 pages showed that 18 pages contained no errors, 12 pages contained just 1 error, 11 pages contained exactly 2 errors, 6 pages contained exactly 3 errors, and the remaining 3 pages contained 4 or more errors.
(a) Present the above information in the form of a table.
(b) Determine the median number of errors per page.
(c) State why it is not possible to calculate the mean number of errors per page.
(d) Find the value of a measure of dispersion.

☐ **LEVEL 2**

8 The frequency distribution in Table 4.19 is based on 83 observations of a discrete variable x.
(a) Calculate, to two decimal places, the mean and the standard deviation of the distribution.
(b) Deduce the mean and the standard deviation of a data set obtained from that given in Table

4.19: (i) by adding 5 to each x-value, (ii) by multiplying each x-value by 2, (iii) by adding 5 to each x-value and multiplying the result by 2.
(b) Determine the median and the 40–60 percentile range of the distribution.

Table 4.19

Value of x	7	8	9	10	11	12	13
Frequency	2	9	15	20	20	13	4

9 A factory has 40 similar machines for producing certain components. Over a particular period of time the number of occasions on which each machine broke down was recorded. It was found that during this period 8 machines did not break down at all, 22 broke down once only, 4 broke down twice, 2 broke down three times, 2 broke down four times, 1 broke down six times, and 1 broke down seven times.
(a) Present the above data in the form of a table and in the form of a diagram.
(b) Name and determine the values of three measures of location of the distribution.
(c) State, with your reason, which of the measures you have named in (b) is inappropriate for the data given.

10 The distribution of the number of customers per day in a departmental store has mean 58 and median 61.
(a) State whether the distribution of the number of customers per day is symmetrical or skew.
(b) Assuming that the distribution of the number of customers per day has a mode, state whether the value of the mode is less than 58, or greater than 61, or between 58 and 61.

11 Table 4.20 shows the frequency distribution of the number of children per family in a sample of 40 families living on a large housing estate.

Table 4.20

No. of children	0	1	2	3
No. of families	3	20	15	2

(a) Calculate the mean, the mean absolute deviation from the mean, and the standard deviation of the distribution.
(b) For another sample of 60 families living on the same housing estate it was found that the mean number of children per family was exactly equal to that for the sample of 40 families, that the mean absolute deviation from the mean was 0·51, and that the standard deviation was 0·30. Calculate the mean absolute deviation from the mean and the standard deviation of the numbers of children per family in all 100 families.

12 A particular brand of saccharine is sold in five differently sized packets which are stated to contain 100, 200, 300, 500, and 700 pellets respectively. Over a period of one month a chemist had sold 30 packets of the saccharine as shown in Table 4.21.

Table 4.21

Stated contents	100	200	300	500	700
No. of packets sold	6	11	6	4	3

(a) Determine the mode, the median, the mean, and the standard deviation of the numbers of pellets per packet sold.
(b) Which of the summary measures given in (a) would have different values if no packets containing 700 pellets were available and 3 additional packets of 500 pellets had been sold instead?
(c) Suppose that each packet actually contained 2% more pellets than was stated on the packet. Write down the values of the mode, the median, the mean, and the standard deviation of the actual numbers of pellets per packet sold.

13 The marks, out of 20, obtained by a class of 80 children in an English test had the distribution shown in Table 4.22.
(a) Calculate the mean and the standard deviation of the marks.
(b) The same children were also given a Mathematics test which was marked out of 50; the marks obtained had a mean of 34·32 and a standard deviation of 4·31. One of the children obtained 16 marks in the English test and 35 marks in the Mathematics test. In which subject did this child have the better performance relative to the whole class?

Table 4.22

Mark (out of 20)	No. of children
10	2
11	3
12	5
13	8
14	9
15	11
16	16
17	14
18	9
19	3

5
GROUPED FREQUENCY DISTRIBUTIONS

"Measure a thousand times and cut once."
Turkish proverb.

In this chapter we consider a large or moderately large data set on a continuous variable.

5.1 Measurement Accuracy

Determining the value of an observation of a discrete variable is easy. In most cases it only involves counting; for example, the number of sweets in a tube of Smarties or the number of children in a family. Determining the value of an observation of a continuous variable is less easy because it usually requires the use of some measuring device which will give a result of limited accuracy (for example, when the variable of interest is length, weight, temperature or time).

Consider a weighing device which will give the weight of an object accurately to the nearest gram only. A recorded weight of 9 g from this weighing device corresponds to an actual weight that may be anywhere in the interval from $9-0.5=8.5$ g to $9+0.5=9.5$ g. Similarly, for a weighing device that gives weights correct to the nearest 0.5 g, a recorded weight of 9.0 g corresponds to an actual weight that may be anywhere in the interval from $9.0-0.25=8.75$ g to $9.0+0.25=9.25$ g. For a weighing device that gives weights correct to the nearest 0.1 g, a recorded weight of 9.0 g corresponds to an actual weight that may be anywhere in the interval from $9.0-0.05=8.95$ g to $9.0+0.05=9.05$ g.

For a different type of example, consider the ages of a group of persons and suppose that each person's age is recorded in completed years only (as is often the case). Then a recorded age of 25, for example, would mean that the person's actual age may be anywhere in the time interval from 25 years up to but not including 26 years.

● **EXERCISE 5.1**

1 A person's height is measured correct to the nearest centimetre. What can you say about the actual height of a person whose recorded height is 175 cm?

2 A weather forecaster states that tomorrow's daytime temperature will be 13°C. Assuming that the forecaster has quoted the temperature to the nearest degree Celsius, what can you say about the actual level of tomorrow's daytime temperature?

3 The weight of an object is stated to be 58.6 kg correct to the nearest tenth of a kilogram. What can you say about the actual weight of the object?

4 The time taken by an athlete to run a race is measured correct to the nearest one-hundredth of a second. Determine the interval of the possible times taken by an athlete for whom the recorded time was 3 minutes and 58.56 seconds.

5 A quartz wrist-watch is programmed to buzz on every hour. When the battery is dead, the time is not displayed on the face of the watch and the watch will not emit a buzz. One day the owner of such a watch heard the buzz at 15.00 hours but there was no buzz at 16.00 hours. Write down an interval of possible times that day when the battery became dead.

5.2 Grouped Data

We shall now consider tabular and graphical representations of data on a continuous variable.

○ **EXAMPLE 5.1**

From Table 1.1, p. 2 the weights of the sweets in 36 tubes of Smarties recorded to the nearest 0.5 g are as follows:

35.0, 35.5, 37.0, 34.0, 36.0, 35.0, 34.5, 36.5, 35.5,
33.5, 35.5, 35.5, 34.0, 36.5, 35.5, 34.0, 35.0, 36.0,
35.0, 35.0, 34.5, 36.0, 36.0, 34.0, 36.5, 35.5, 36.0,
36.0, 34.5, 35.0, 35.0, 36.5, 35.5, 37.0, 34.5, 35.0.

To simplify the presentation of these data let us first construct a tally chart to determine the frequency distribution of the recorded weights. The chart is displayed in Table 5.1.

Since each weight was recorded to the nearest 0.5 g, the lowest recorded weight of 33.5 g corre-

Table 5.1 *Tally Chart for the Weights of Sweets in 36 Tubes of Smarties*

Recorded weight (g)	Tally	Frequency
33.5	I	1
34.0	IIII	4
34.5	IIII	4
35.0	JHT III	8
35.5	JHT II	7
36.0	JHT I	6
36.5	IIII	4
37.0	II	2
	Total	36

sponds to an actual weight that may be anywhere in the interval from $33.5 - 0.25 = 33.25$ g to $33.5 + 0.25 = 33.75$ g, while the next lowest recorded weight of 34.0 g corresponds to an actual weight that may be anywhere in the interval from 33.75 g to 34.25 g. Since both of these intervals include the value 33.75, it is not clear whether an actual weight of 33.75 g is in the first or the second interval. To avoid such an ambiguity we shall adopt the convention that any quoted interval for actual values consists of the values from the lower limit of the interval up to but *not including* the upper limit of the interval. Thus in the case under consideration, an actual weight of 33.75 g will be included in the second of the two intervals only.

Replacing each recorded weight in Table 5.1 by its corresponding interval of actual weights, the distribution of the actual weights of the sweet in the 36

Table 5.2 *Grouped Frequency Distribution of the Actual Weights of the Sweets in 36 Tubes of Smarties*

Actual weight (g)	Frequency
33.25–	1
33.75–	4
34.25–	4
34.75–	8
35.25–	7
35.75–	6
36.25–	4
36.75–	2
37.25–	0

tubes may be displayed as in Table 5.2, which is referred to as a **grouped frequency distribution table**, the intervals (or groups) of values being referred to as the **class intervals**. Observe that Table 5.2 includes the extra interval '37.25–', having frequency zero, in order to make it clear that the penultimate interval '36.75–' is from 36.75 up to but not including 37.25. An alternative, equally acceptable representation of the class intervals is in terms of the symbols ⩾ (denoting 'greater than or equal to' or 'at least') and < (denoting 'less than'). Using these symbols, the top row in Table 5.2 would appear as:

Actual weight (g)	
⩾	<
33.25	33.75
33.75	34.25
34.25	34.75
34.75	35.25
35.25	35.25
35.75	36.25
36.25	36.75
36.75	37.25

A grouped frequency distribution may be exhibited diagrammatically using the same principles as those for bar diagrams described in Section 2.2, p. 8. The main difference is that now the variable axis will be *scaled*. The aim is to draw rectangles on the class intervals having *areas* proportional to the corresponding frequencies. Since in our present example all the class intervals have the same width (namely 0.5 g), drawing rectangles having heights proportional to the frequencies will produce rectangles whose areas are also proportional to the frequencies. (The case of unequal class intervals will be discussed in Section 5.7.) The resulting diagram is referred to as a **histogram**. The histogram for the grouped frequency distribution in Table 5.2 is shown in Figure 5.1. In this diagram we have marked the mid-values of the class intervals along the variable axis instead of marking both limits of each class, as this avoids clutter.

It often happens that data collected on a continuous variable are such that the number of distinct recorded values that occur is so large that a frequency distribution of the recorded values will be too cumbersome to give a picture of the pattern of the distribution. In such circumstances it is advisable to group the recorded values into a moderate number of intervals. The number of intervals chosen should not be so small that it results in a substantial loss of information, and should not be so large that there is no distributional pattern apparent. Between 5 and 10 intervals is usually about right in most cases, but the precise number of intervals to use and the values of their limits is often a matter of trial and error until a suitable choice is found.

○ EXAMPLE 5.2

From Table 1.2, p. 3 the weights, measured to the nearest 0.1 kg, of 20 boys in a class were as follows:

54.8, 59.5, 62.2, 59.9, 66.7,
64.7, 56.3, 58.1, 62.2, 58.8,
65.4, 59.0, 59.5, 56.8, 57.9,
64.9, 61.3, 68.6, 68.4, 62.9.

(1) Construct a tally chart for the grouped frequency distribution of these recorded weights using the five class intervals:

54.8–57.5, 57.6–60.3, 60.4–63.1, 63.2–65.9, 66.0–68.7.

(2) Set up the corresponding grouped frequency distribution of the actual weights of the 20 pupils and exhibit it as a histogram.

(1) The tally chart for the specified class intervals is shown in Table 5.3. The first recorded weight, 54.8 kg, falls in the first of the intervals, the second

Figure 5.1 *Histogram for the Data in Table 5.2*

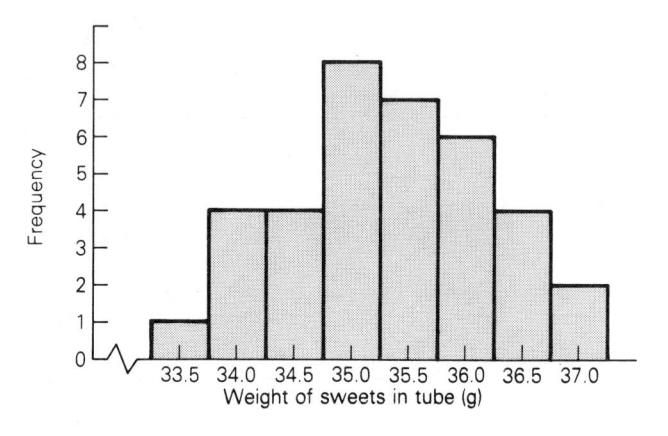

Table 5.3 *Tally Chart for the Grouped Frequency Distribution of the Recorded Weights of 20 Boys*

Recorded weight (g)	Tally	Frequency
54.8–57.5	III	3
57.6–60.3	JHT II	7
60.4–63.1	IIII	4
63.2–65.9	III	3
66.0–68.7	III	3
	Total	36

recorded weight of 59.5 kg falls in the second interval, and so on for all 20 recorded weights, a vertical stroke being placed alongside the class interval in which a recorded weight falls.

(2) Since each weight was recorded to the nearest 0.1 kg it follows, for example, that recorded weights in the interval 54.8–57.5 cover actual weights ranging from 54.75 up to but not including 57.55, and recorded weights in the interval 57.6–60.3 cover actual weights ranging from 57.55 up to but not including 60.35. Modifying all the intervals in Table 5.3 similarly, the grouped frequency distribution of the actual weights is as displayed in Table 5.4, where again we have included a final interval having frequency zero so as to avoid any possible misinterpretation of the upper limit of the penultimate interval.

Table 5.4 *Grouped Frequency Distribution of the Actual Weights of 20 Boys*

Actual weight (g)	Frequency
54.75–	3
57.55–	7
60.35–	4
63.15–	3
65.95–	3
68.75–	0

The corresponding histogram is shown in Figure 5.2. Observe that since all the class intervals have the same width (2.8 kg), it is sufficient to draw rectangles having heights equal to the frequencies.

Figure 5.2 *Histogram of the Actual Weights of 20 Boys*

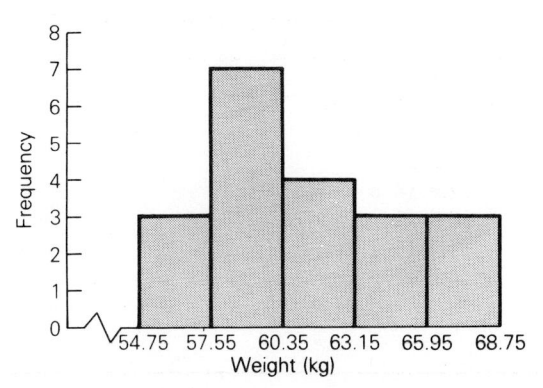

An alternative diagrammatic representation of a grouped frequency distribution is the **frequency polygon**, in which the class frequencies are plotted against the mid-values of the class intervals and joined by straight lines. The frequency polygon for the data in Table 5.4 is displayed in Figure 5.3.

Figure 5.3 *Frequency Polygon for the Data in Table 5.4*

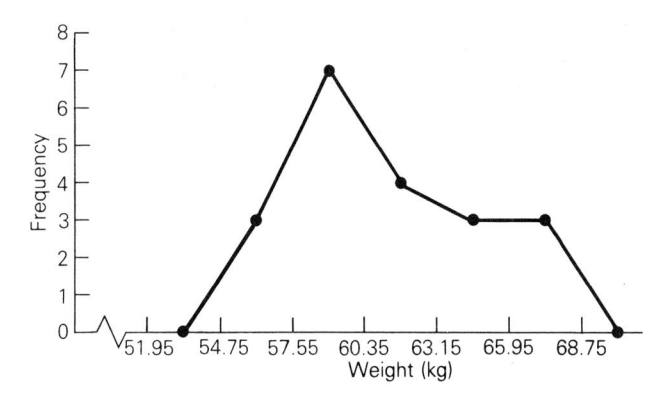

DISTRIBUTION SHAPES

As was the case for discrete frequency distributions (see Section 4.5, p. 43) grouped frequency distributions arise in various shapes. Some of these shapes are displayed in Figure 5.4 where, for convenience, the frequency polygons have been replaced by smooth curves.

One of the most important distributions for a continuous variable is known as the **Normal distribution**. Its shape is shown in Figure 5.4a. This distribution is symmetrical about its unique mode,

Figure 5.4 *Some Shapes of Grouped Frequency Distributions: (a) Symmetrical and Unimodal, (b) Bimodal, (c) Positively Skew, (d) Negatively Skew*

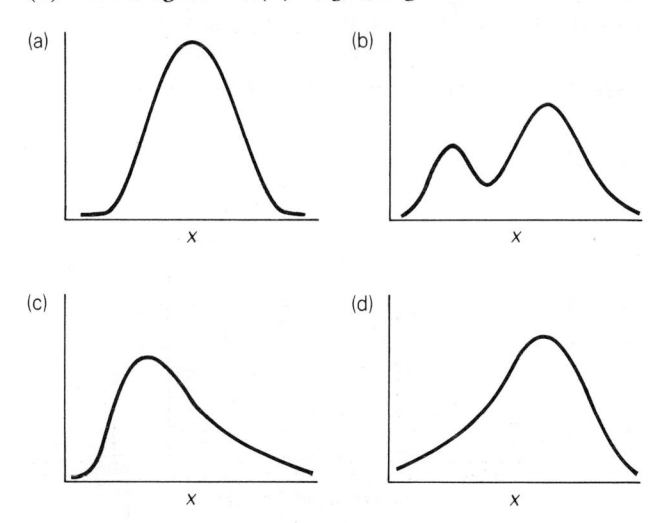

[51

which is also the mean (and the median) of the distribution. The spread (dispersion) of a Normal distribution is measured by its standard deviation. If two Normal distributions have different means but equal standard deviations then the two distributions will have identical shapes. Diagrammatically, the one with the greater mean will be to the right along the x-axis of the one with the smaller mean. If two Normal distributions have equal means but unequal standard deviations then diagrammatically they will both be centred about the same point, the one with the larger standard deviation being 'fatter' (more spread out). Some examples of continuous variables that have been found to be approximately Normally distributed include (1) the heights of adult males (or females), (2) examination marks, and (3) errors in repeated measurements.

A bimodal distribution, as illustrated in Figure 5.4b may occur as the result of two distributions having been combined into one distribution. For example, the distribution of the heights of adults (both males and females) would be bimodal, the mean height of adult males being greater that the mean height of females. Thus, with reference to Figure 5.4b, the right hand part will consist mainly of heights of males and the left hand part will consist mainly of heights of females. (To demonstrate this, construct a grouped frequency distribution and corresponding frequency polygon or histogram of the heights of all 36 pupils given in Table 1.2, p. 3.)

A skew distribution, as illustrated in Figures 5.4c and d, is one in which the mean, the median and mode (if it exists) have different values. This occurs when the data set includes some extreme values, much larger or much smaller than the majority of the values. For example, data on incomes and data on ages at marriage have positively skewed distributions, while data on ages at death from a heart attack have a negatively skewed distribution. The frequency polygon of the weights of 20 boys given in Figure 5.3 is seen to be positively skew, the 'tail' on the right being somewhat longer than the 'tail' on the left.

● EXERCISE 5.2

1 Table 1.2, p. 3 gives the heights of 20 boys to the nearest centimetre. Using the class intervals 159–161, 162–164, 165–167, 168–170, 171–173, 174–176 for the recorded heights, construct a grouped frequency distribution table for the boys' actual heights and present them as a histogram. Retain your table for use in later exercises.

2 In tests on the fuel consumption of a particular model of car under urban driving conditions the following consumption figures, in kilometres per litre of fuel, to the nearest tenth of a kilometre, were obtained in 40 tests.

9.8, 11.4, 12.7, 10.8, 13.1, 11.5, 11.0, 12.4,
10.3, 11.9, 10.5, 12.3, 11.4, 11.8, 12.9, 10.4,
12.2, 11.1, 11.6, 12.4, 11.2, 10.1, 11.8, 12.1,
11.3, 12.8, 11.9, 10.7, 12.0, 11.6, 11.5, 12.3,
10.7, 11.9, 12.5, 9.9, 11.8, 12.6, 11.8, 12.0.

Construct a grouped frequency distribution table for the actual consumptions in the 40 tests using seven equal class intervals. The first interval for the recorded consumptions should be 9.8–10.2. Present the distribution as a histogram. Retain your table for use in later exercises.

3 A grouped frequency distribution of the lifetimes, in hundreds of hours, of 60 electric light bulbs of a particular wattage and brand is given in Table 5.5. Present this distribution as a histogram and comment on its shape.

Table 5.5

Lifetime ('00 hours)	No. of bulbs
8–	4
9–	8
10–	21
11–	9
12–	7
13–	5
14–	3
15–	2
16–	1
17–	0

4 Table 5.6 shows a grouped frequency distribution of the ages at marriage in complete years, of 100 married males. Present this distribution as a histogram and comment on its shape.

Table 5.6

Age at marriage (years)	No. of males
16–21	14
22–27	29
28–33	24
34–39	15
40–45	8
46–51	5
52–57	3
58–63	2

It may happen that data on a discrete variable consists of such a large number of distinct values that it will be difficult to see any pattern in the frequency distribution table. In such circumstances the distinct values may be grouped together appropriately so as to provide a meaningful summary of the data in the form of a grouped frequency distribution table. As an example, consider an examination paper which is marked out of 100. There may be up to 101 different marks (0, 1, 2, ..., 100) on such a paper. One possible grouping of such marks could be:

1–20, 21–40, 41–60, 61–80, 81–100,

consisting of five class intervals each of width 20 marks; we have assumed here that no candidate scored zero in the examination. Assuming that the marks were allocated on a continuous scale of measurement and that each mark awarded was then rounded to the nearest integer, the grouped frequency distribution of the marks may be displayed as a histogram. This would mean treating the mark-group 1–20 as corresponding to actual marks from 0.5 up to but not including 20.5, with similar interpretations of the other mark-groups. (See Question 5 below for a specific example.)

Another example of a discrete variable which is often more conveniently treated as a continuous variable is one measured in monetary units. Table 5.7 shows a grouped frequency distribution of the earnings in a year of 72 insurance salesmen. The first earnings-group '10–' corresponds to earnings ranging from £10 000 to £10 999.99. However, it is far more convenient to interpret the interval as ranging from £10 000 up to but not including £11 000, as though money were measured on a continuous scale. (Question 6 of Exercise 5.2 asks you to draw the histogram for the data in Table 5.7.)

Table 5.7 *Earnings in a year of 72 Insurance Salesmen*

Earnings (£'000)	No. of salesmen
10–	2
11–	6
12–	12
13–	18
14–	16
15–	10
16–	8
17–	0

5 A distribution of the marks out of a maximum of 100 obtained in an examination taken by 180 candidates is displayed in Table 5.8. Present this

Table 5.8

Marks	Frequency
1–10	3
11–20	12
21–30	15
31–40	25
41–50	42
51–60	36
61–70	27
71–80	11
81–90	6
91–100	3

distribution as a histogram and comment on its shape.

6 *Draw a histogram of the yearly earnings of 72 insurance salesmen as displayed in Table 5.7.*

5.3 Modal Class

You may recall that the mode of a data set is defined to be the value that occurs most frequently. We saw in Section 4.2 that for data on a discrete variable in the form of a frequency distribution, the existence and the value of the mode may be obtained directly from the frequency distribution table. When the variable is continuous and a frequency distribution table of the *recorded* values of the variable is constructed, then again the existence and the value of the mode of the recorded values can be obtained directly from the table. However, since the *actual* values of the observations are not available to us there is no way in which we can determine the mode of the actual values.

For a grouped frequency distribution, the class interval (if there is one) of the actual values having the highest frequency is called the **modal class**. For example:

in Table 5.2 the modal class is 34.75–35.25 with frequency 8,
in Table 5.4 the modal class is 57.55–60.35 with frequency 7,
in Table 5.7 the modal class is 13–14 with frequency 18

Note that there is no guarantee at all that the mode

[53

of the actual values is in the modal class; any attempt at trying to find the mode of the actual values is futile.

● **EXERCISE 5.3**

Questions **1–6** Write down the modal class of each of the grouped frequency distributions in Exercise 5.2.

5.4 Cumulative Frequency Diagrams

For a frequency distribution the cumulative frequency of any distinct value of the variable is the total frequency of the observations less than or equal to that value (see Section 4.3, p. 39). We now extend this notion to a grouped frequency distribution.

○ **EXAMPLE 5.3**

Consider again the grouped frequency distribution of the actual weights of the sweets in 36 tubes of Smarties displayed in Table 5.2 and reproduced here for ease of reference.

Grouped Frequency Distribution of the Actual Weights of the Sweets in 36 Tubes of Smarties

Actual weight (g)	Frequency
33.25–	1
33.75–	4
34.25–	4
34.75–	8
35.25–	7
35.75–	6
36.25–	4
36.75–	2

From this table we see that:

1 tube contained sweets weighing under 33.75 g,
1+4 = 5 tubes contained sweets weighing under 34.25 g,
5+4 = 9 tubes contained sweets weighing under 34.75 g,

and so on. The figures on the left are the **cumulative frequencies** corresponding to the weights given on the right. When dealing with a grouped frequency distribution it is most important to note that the cumulative frequencies are associated with variable values that are the *upper limits* of the class intervals. The cumulative frequency distribution corresponding to the data in this table is displayed in Table 5.9.

Table 5.9 *Cumulative Frequency Distribution of the Actual Weights of Sweets in 36 Tubes of Smarties*

Upper weight limit (g)	Cumulative frequency
33.75	1
34.25	5
34.75	9
35.25	17
35.75	24
36.25	30
36.75	34
37.25	36

Let us now graph this distribution. Having drawn and scaled the variable and cumulative frequency axes, we plot the cumulative frequencies (CF) against the corresponding upper limits of the weight class intervals. Then we join up the plotted points with *straight lines*; the graph drawn is known as a **cumulative frequency polygon**, or sometimes as an **ogive**. The cumulative frequency polygon for the data in Table 5.9 is displayed in Figure 5.5. Observe that we have also plotted the point (33.25, 0) on the graph, since we know that no tube (CF = 0) contained sweets weighing less than 33.25 g.

Figure 5.5 *Cumulative Frequency Polygon for the Data in Table 5.9*

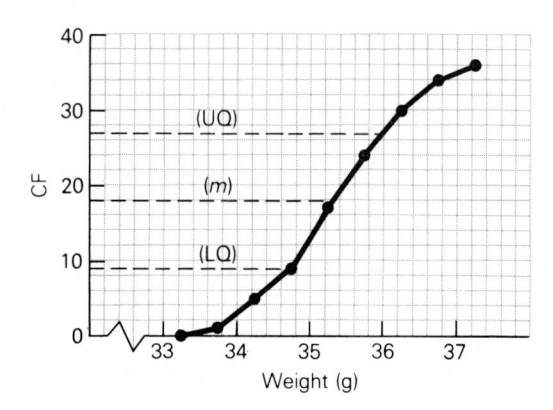

Joining up the plotted points with straight lines is justified here by making the assumption that the observations in any class interval are uniformly spread throughout that interval. That is, as we proceed through a class interval the cumulative frequency will increase at a steady rate. Contrast this with the situation when the variable is discrete (see Section 4.3, p. 39).

Since our intention is to use the cumulative frequency graph to find variable values corresponding to specified values of the cumulative frequency, and vice versa, it is important that we have easy-to-read scales along both axes. In Figure 5.5 we have scaled the vertical axis so that a distance of 1 cm represents 10 units of cumulative frequency, and the horizontal axis so that 1 cm represents 1.0 g weight. Thus 1 mm along the vertical axis represents 1 tube and 1 mm along the horizontal axis represents 0.1 g.

The cumulative frequency polygon provides a means for determining *estimates* of the median and the quartiles (and indeed any other percentiles) of the distribution. Only estimates can be made (as opposed to the true values obtained from the cumulative frequency graphs in Section 4.3) because only the recorded values (as opposed to the actual values) are available to us. The estimates should be reasonably close to the true values provided that our assumption (justifying straight line joins of the plotted points) is strictly valid.

For an estimate of the median of the weights we need the variable value for which $CF = \frac{1}{2} \times 36 = 18$. From Figure 5.5 our estimate of the median, is:

$$m = 35.30 \text{ g}.$$

For estimates of the lower quartile LQ and the upper quartile UQ we need the variable values for which $CF = \frac{1}{4} \times 36 = 9$ and $CF = \frac{3}{4} \times 36 = 27$ respectively. Reading from the graph in Figure 5.5 we find that the required estimates are:

$$LQ = 34.75 \text{ g},$$
$$UQ = 36.00 \text{ g}.$$

(Referring back to Figure 5.1, the histogram of the actual weights of the sweets, if vertical lines are drawn from the values of the quartiles (including the median) to meet the horizontal bars of the histogram then these lines divide the total area under the histogram into four equal parts. In particular, the line from the median bisects the area of the histogram.)

Our estimate of the semi-interquartile range of the distribution of the weights of the sweets is thus:

$$SIQR = \tfrac{1}{2}(36.00 - 34.75) = 0.625.$$

A summary of the distribution of the weights is

Figure 5.6 *Box-Whisker Plot for the Data in Table 5.9*

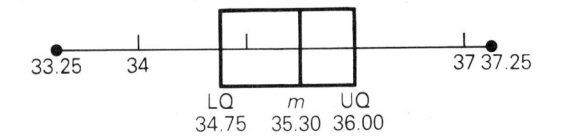

displayed in the form of a box-and-whisker plot in Figure 5.6.

As discussed on p. 41 we could have drawn the **cumulative relative frequency polygon** if we so wished. The cumulative relative frequencies are obtained by dividing the cumulative frequencies by the total number of observations in the data set; they are associated with the upper limits of the class intervals. In this example we would divide the cumulative frequencies by 36; the resulting cumulative relative frequencies would then be plotted against the upper limits of the class intervals and we would join the plotted points with straight lines.

○ EXAMPLE 5.4

Table 5.10 shows a grouped frequency distribution of the ages, in years, of the males killed in road accidents in Camfordshire in 1986.

Table 5.10 *Grouped Frequency Distribution of the Ages of Males Killed in Road Accidents in Camfordshire in 1986*

Age (years)	No. of males killed
0–19	22
20–39	28
40–59	19
60–79	6
80–99	3
Total	78

Proceeding as in Example 5.3, the cumulative frequencies are obtained by adding the frequencies in the table successively; they are to be associated with the upper limits of the successive class intervals. The cumulative relative frequencies are then obtained by dividing the cumulative frequencies by 78, the total number of observations in this example. The upper limits of the class intervals, the cumulative frequencies (CF) and the cumulative relative frequencies (CRF) to 3 decimal places, are displayed in Table 5.11.

Table 5.11 *Cumulative Relative Frequency Distribution from Table 5.10*

Upper age limit (years)	CF	CRF
20	22	0.282
40	50	0.641
60	69	0.885
80	75	0.962
100	78	1.000

Since the cumulative relative frequencies extend to the value 1, when drawing the cumulative relative frequency polygon we shall scale the vertical axis from 0 to 1. The cumulative relative frequency polygon for our present example is shown in Figure 5.7. The scale along the variable (horizontal) axis is 1 cm to represent 20 years, and the scale along the CRF (vertical) axis is 1 cm to represent 0.2.

Figure 5.7 *Cumulative Relative Polygon from Table 5.11*

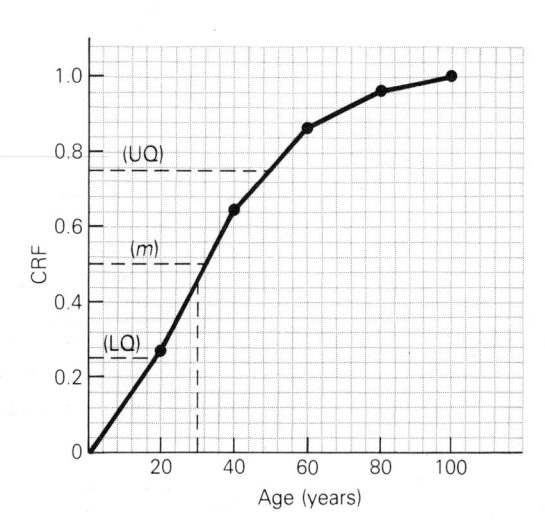

To estimate the median age m we need the variable value for which CRF = 0.5; reading from the graph in Figure 5.7 our estimate is:

$$m = 32.0 \text{ years.}$$

To estimate the quartiles we need the variable values for which CRF = 0.25 and CRF = 0.75; reading from the graph our estimates of the lower quartile (LQ) and the upper quartile (UQ) are respectively:

LQ = 18.0 years,

and

UQ = 49.0 years.

Similarly, if we required an estimate of the 60th percentile of the distribution we would read from the graph the variable value for which CRF = 0.6; doing so, our estimate is seen to be $P_{60} = 37.5$ years.

The polygon may also be used to estimate the proportion of males killed who were under some specified age. For example, an estimate of the proportion of the males killed who were under 30 years old is given by the CRF for which the variable value is 30; from the polygon we find that the estimate of the proportion is 0·46 or 46%.

● EXERCISE 5.4

1 For the distribution of the ages at marriage of 100 males given in Table 5.6, p. 52:
(a) construct the cumulative frequency table,
(b) draw the cumulative frequency polygon,
(c) estimate the median, the lower quartile, the upper quartile, and the 40th and 60th percentiles,
(d) estimate the percentage of the males who got married when they were under 30 years of age.

2 For the distribution of the lifetimes of 60 electric light bulbs given in Table 5.5, p. 52:
(a) draw the cumulative relative frequency polygon,
(b) estimate the median and the quartiles,
(c) estimate the percentage of the bulbs that had a lifetime of 1260 hours or more,
(d) *summarize the distribution in the form of a box-and-whisker plot.

3 For the distribution of marks given in Table 5.8, p. 53:
(a) estimate the median and the quartiles,
(b) estimate the percentage of the candidates whose marks were in the range from 55 to 74.

5.5 Percentiles Using Linear Interpolation*

Constructing a graph of the cumulative (relative) frequency polygon for a grouped frequency distribution is not essential for the estimation of percentiles. An alternative method, based on linear interpolation, is illustrated in the following examples.

○ **EXAMPLE 5.5**

Consider the grouped frequency distribution of the actual weights of the sweets in 36 tubes of Smarties given in Table 5.2. The corresponding cumulative frequency distribution is displayed in Table 5.9 and is reproduced here for convenience. To estimate the median we need the variable value for which $CF = \frac{1}{2} \times 36 = 18$. From the table we can see that the median is in the interval from 35.25 to 35.75. In this interval the cumulative frequency increases by $(24-17) = 7$. We shall assume that the CF increases throughout this interval at a steady rate, which is precisely the assumption we made to justify the straight line joins of the plotted points when drawing the cumulative frequency polygon (Figure 5.5).

Cumulative Frequency Distribution from Table 5.9

Upper weight limit (g)	CF
33.75	1
34.25	5
34.75	9
35.25	17
35.75	24
36.25	30
36.75	34
37.25	36

Since the width of the interval 35.25–35.75 is 0.5 g, this assumption means that the cumulative frequency increases throughout the interval at a rate of 7 per 0.5 g, or equivalently, at the rate of 1 per $(0.5/7)$ g. The median is the value in the interval at which the cumulative frequency reaches the value 18, which means it has increased by $18 - 17 = 1$ from its value at the lower limit of the interval. Thus the corresponding increase in the weight for the cumulative frequency to reach the value 18 is:

$$1 \times \frac{0.5}{7} = 0.07 \quad \text{(to 2 d.p.)}.$$

Hence our estimate of the median weight is:

$$m = 35.25 + 0.07 = 35.32 \quad \text{(to 2 d.p.)},$$

as compared with our graphical estimate of 35.30 obtained in the preceding section; the discrepancy is due to the accuracy limitations when reading from a graph.

Now consider the estimation of the lower quartile, which is the variable value for which $CF = \frac{1}{4} \times 36 = 9$. From the table we see that the lower quartile LQ is the last observation in the interval 34.25–34.75, and it therefore has the value 34.75, agreeing with the graphical estimate we obtained in Section 5.4.

For an estimate of the upper quartile UQ we need the variable value for which $CF = \frac{3}{4} \times 36 = 27$. From the table we see that UQ is in the interval 35.75–36.25, in which CF is increasing at the rate of $(30-24) = 6$ per 0.5 g or 1 per $(0.5/6)$ g. The value of UQ corresponds to an increase in CF of $(27-24) = 3$, so that our estimate of UQ is given by:

$$UQ = 35.75 + 3 \times \frac{0.5}{6} = 36.00 \text{ g} \quad \text{(to 2 d.p.)},$$

agreeing exactly with our graphical estimate in Section 5.4.

○ **EXAMPLE 5.6**

Here is the cumulative frequency distribution of the ages of males killed in road accidents in Camfordshire in 1986 (from Table 5.10).

Since the total number of observations is 78,

Cumulative Frequency Distribution of the Ages of Males Killed in Road Accidents in Camfordshire in 1986

Upper age limit (years)	CF
20	22
40	50
60	69
80	75
100	78

estimates of the median, lower quartile and upper quartile can be obtained from the variable values for which CF is 39, 19.5, and 58.5 respectively.

From the table, the median, being the variable value for which $CF = 39$, is seen to be in the interval 20–40. This interval is of width 20 and the cumulative frequency increases through the interval by $(50-22) = 28$, corresponding to a steady rate increase of 1 per $(20/28)$ years. For the median, the CF has to increase by $(39-22) = 17$. Hence our estimate of the median age is given by:

$$m = 20 + 17 \times \frac{20}{28} = 32.14 \text{ years} \quad \text{(to 2 d.p.)}.$$

From the table, the lower quartile, corresponding to $CF = 19.5$, is seen to be in the interval 0–20. In this interval CF increases from 0 to 22, equivalent to a steady rate of increase equal to 1 per $(20/22)$ years. Thus, our estimate of the lower quartile LQ is:

$$LQ = 0 + 19.5 \times \frac{20}{22} = 17.73 \text{ years} \quad \text{(to 2 d.p.)}.$$

Finally the upper quartile, corresponding to

CF = 58.5, is seen to lie in the interval 40–60. This interval is also of width 20, and the cumulative frequency increases through the interval by (69−50) = 19, equivalent to a steady increase of 1 per (20/19) years. The upper quartile UQ is the variable value for which CF = 58.5, corresponding to an increase in CF of (58.5−50) = 8.5 from its value at the lower limit of the interval 40–60. Hence our estimate of UQ is given by:

$$UQ = 40 + 8.5 \times \frac{20}{19} = 48.95 \text{ years} \quad \text{(to 2 d.p.)}.$$

The estimates obtained here differ slightly from those obtained graphically in Example 5.4 because of the limited accuracy when reading from a graph.

Now consider how we can estimate the proportion of the males killed who were under 30 years old. From the table we see that in the interval 20–40 the cumulative frequency increased by (50−22) = 28, corresponding to a steady increase of 28 per 20 years. Since 30 is 10 more than 20, the cumulative frequency corresponding to an age of 30 is given by:

$$22 + 28 \times \frac{10}{20} = 36.$$

Thus our estimate of the *number* of males killed who were under 30 years old is 36. The corresponding estimate of the *proportion* of males killed who were under 30 years old is 36/78 = 0.46 to two decimal places, or 46% to the nearest whole number; this agrees with our graphical estimate in Example 5.4.

● **EXERCISE 5.5**

Questions **1–3** *Use the method of linear interpolation to obtain the answers to Questions 1(c), (d); 2(b), (c), and 3(a), (b) of Exercise 5.4 p. 56.*

5.6 Estimating the Mean and Standard Deviation

To estimate the mean and the standard deviation of a data set which is in the form of a grouped frequency distribution we replace each class interval with its mid-value and apply the procedures described in Chapter 4. The mid-value of a class interval is the average of its lower and upper limits. This effectively replaces the grouped frequency distribution with a frequency distribution in which the variable values are the mid-values of the class intervals. The assumption being made to justify this procedure is that the observations in any class interval are fairly evenly spread throughout that interval.

○ **EXAMPLE 5.7**

Consider again the weights of the sweets in 36 tubes of Smarties. The grouped frequency distribution of the actual weights is displayed in Table 5.2. The mid-values of the class intervals in this case are in fact, the recorded weights as displayed in Table 5.1. For

Table 5.12 *Frequency Distribution of the Mid-interval Values of the Weights of Sweets in 36 Tubes of Smarties*

Class interval	Mid-value (x)	No. of tubes (f)
33.25–	33.5	1
33.75–	34.0	4
34.25–	34.5	4
34.75–	35.0	8
35.25–	35.5	7
35.75–	36.0	6
36.25–	36.5	4
35.75–	37.0	2

convenience, they have been included in Table 5.12. As explained in Chapter 4, the arithmetic involved when calculating a mean or standard deviation of a frequency distribution may often be reduced substantially by choosing an appropriate working origin and working scale. (that is, by using a transformation). Let x denote the mid-value of a class interval. Subtracting 33.5 from each x-value in Table 5.12 will yield the simpler values: 0, 0.5, 1, 1.5, 2, 2.5, 3, 3.5, respectively. On multiplying each of these values by 2 we get the integer values:

$$0, 1, 2, 3, 4, 5, 6, 7,$$

respectively. These operations are equivalent to applying the transformation:

$$y = 2(x - 33.5).$$

The frequency distribution of the y-values is shown in Table 5.13.

Table 5.13 *Frequency Distribution of $y = 2(x - 33.5)$*

y-value	Frequency (f)
0	1
1	4
2	4
3	8
4	7
5	6
6	4
7	2
Total	36

From this table we find that:

$$n = \Sigma f = 36, \quad \Sigma fy = 132, \quad \text{and} \quad \Sigma fy^2 = 596.$$

It follows that the mean of the y-values is given by:

$$\bar{y} = \frac{132}{36},$$

and the mean, from (3.4) on p. 27, of the x-values, namely our estimate of the mean weight of sweets per tube, is:

$$\bar{x} = 33.5 + \tfrac{1}{2}\bar{y} = 33.5 + \frac{132}{72} = 35.33 \text{ g} \quad \text{(to d.p.)}.$$

From (4.1) on p. 36 the standard deviation of the y-values is given by:

$$\text{SD}(y) = \frac{1}{n}\sqrt{(n\Sigma fy^2 - (\Sigma fy)^2)} = \frac{\sqrt{(36 \times 596 - 132^2)}}{36}$$

$$= \frac{\sqrt{4032}}{36}.$$

Since $y = 2(x - 33.5)$ it follows from (3.4) on p. 27 that the standard deviation of the x-values, which is our estimate of the standard deviation of the actual weights of the sweets, is given by:

$$\text{SD}(x) = \tfrac{1}{2}\text{SD}(y) = \frac{\sqrt{4032}}{72} = 0.8819 \quad \text{(to 4 d.p.)}.$$

○ **EXAMPLE 5.8**

Table 5.10 on p. 55 shows a grouped frequency distribution of the ages of males killed in road accidents in Camfordshire in 1986. The age-groups for the ages in complete years are:

$$0-19, 20-39, 40-59, 60-79, 80-99.$$

The class interval for the actual ages in the age-group 0–19 is from 0 up to 20, and its mid-value is $\tfrac{1}{2}(0 + 20) = 10$. Similarly, the mid-values

Table 5.14 *Frequency Distribution of the Mid-values of the Age Class Intervals of Males Killed in Road Accidents in Camfordshire in 1986*

Age group (years)	Mid-value (x)	No. killed
0–19	10	22
20–39	30	28
40–59	50	19
60–79	70	6
80–99	90	3
Total		78

of the other class intervals for actual ages are $\frac{1}{2}(20+40)=30$, $\frac{1}{2}(40+60)=50$, $\frac{1}{2}(60+80)=70$, and $\frac{1}{2}(80+100)=90$ respectively. Thus the frequency distribution of the mid-values of the age class intervals is as shown in Table 5.14.

Since every mid-value is an integer, a transformation is not really essential here. Nevertheless, some simplification of the arithmetic will result from using the transformation $y=(x-10)/20$, which yields the y-values 0, 1, 2, 3, 4 respectively. The frequency distribution of the y-values produced by using this transformation is displayed in Table 5.15.

Table 5.15 *Frequency Distribution of* $y=(x-10)/20$ *for the Data in Table 5.14*

y-value	0	1	2	3	4	Total
Frequency (f)	22	28	19	6	3	78

From the entries in this table we find that:

$n=\Sigma f=78$, $\quad \Sigma fy=96$, $\quad$ and $\quad \Sigma fy^2=206$.

Hence, the mean and the standard deviation of the y-values are:

$$\bar{y}=\frac{1}{n}\Sigma fy=\frac{96}{78},$$

and:

$$SD(y)=\frac{1}{n}\sqrt{(n\Sigma fy^2-(\Sigma fy)^2)}=\frac{\sqrt{(78\times 206-96^2)}}{78}$$
$$=\frac{\sqrt{6852}}{78}.$$

Transforming back to the x-values, our estimates of the mean and the standard deviation of the ages are:

$$\bar{x}=10+20\bar{y}=10+20\times\frac{96}{78}=34.62 \text{ years}$$

(to 2 d.p.),

and:

$$SD(x)=20\ SD(y)=20\times\frac{\sqrt{6852}}{78}=21.22 \text{ years}$$

(to 2 d.p.).

● **EXERCISE 5.6**

Questions *1–6* *Estimate, to two decimal places, the mean and the standard deviation of each of the grouped frequency distributions in Exercise 5.2 on p. 52.*

7 Fifty runners competed in a cross-country race. Table 5.16 shows the numbers of runners who had completed the course by certain times after the start of the race. (Note that this is a *cumulative* frequency distribution table.)

Table 5.16

Time from the start (min)	No. who had completed course
60	0
65	5
70	16
75	24
80	46
85	50

(a) Draw up a grouped frequency distribution table for the times taken by the runners to complete the course.
(b) Estimate the mean and the standard deviation of the times taken, giving each answer to two decimal places.

5.7 Groupings That Have Unequal Class Intervals*

Situations arise where for various reasons it is more informative to summarize a data set in the form of a grouped frequency distribution whose class intervals are not all of the same width. Many of the grouped frequency distributions appearing in government publications, especially those involving people's ages, fall into this category. The procedures described earlier in this chapter are still applicable with little modification, apart from the histogram representation of the distribution.

○ **EXAMPLE 5.9**

Table 5.17 shows a grouped frequency distribution of the ages, in complete years, of 300 females in a certain town in 1986. Draw a histogram to show this distribution.

The essential difference between the grouped frequency distribution in Table 5.17 and those we have met previously is that the class intervals are not all of the same width. For instance, the first interval 0–4 represents actual ages from 0 up to but not including 5 and is of width 5 years, whereas the second interval 5–14 extends from 5 up to but not including 15 and is of width 10 years. Similarly, we find that the third, fourth, and sixth intervals are of

Table 5.17 *Age Distribution of 300 Females*

Age (years)	No. of females
0–4	17
5–14	43
15–29	60
30–44	56
45–64	68
65–79	39
80–99	17
	Total 300

width 15 years, while the fifth and seventh (final) intervals are of width 20 years.

In Section 5.2 we made the point that the histogram for a grouped frequency distribution should have rectangles whose *areas* are proportional to the frequencies they represent. However, since the class intervals we considered were all of the same width, drawing rectangles with heights proportional to the frequencies actually gave rectangles whose areas were also proportional to the frequencies. The situation is different in our present example since the class intervals do not all have the same width.

Figure 5.8 *Incorrect Diagram for Table 5.17*

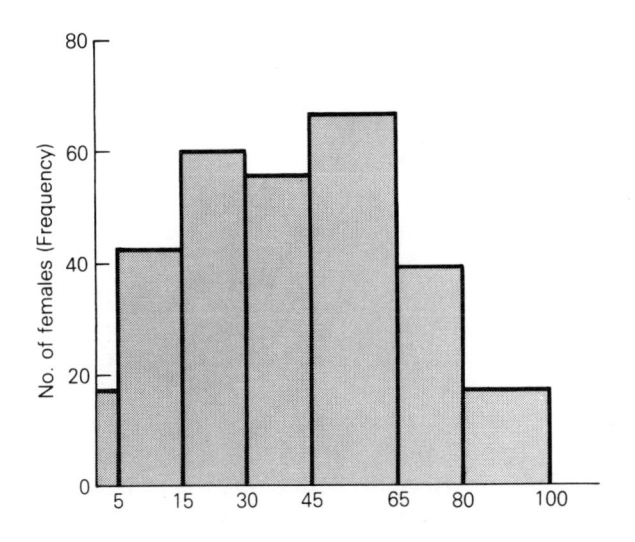

In Figure 5.8 we have drawn a diagram of our grouped frequency distribution, with rectangles having heights proportional to the frequencies. To see how misleading this diagram is, consider the rectangles above the class intervals 0–5 and 80–100. These rectangles are the same height (17 on the frequency scale) but the area of the rectangle above 80–100 is four times that above 0–4; this implies that there are four times as many females in the age-group 80–100 as in the age-group 0–5. Other such false comparisons can be made from Figure 5.8.

We need to ensure that the rectangles we draw above the class intervals have areas which are proportional to the frequencies they represent. To achieve this we modify the vertical axis so that it represents *frequency per one-year of age* rather than frequency. In general, the vertical axis should be scaled in terms of *frequency per unit value* of the variable, which is usually referred to as the **frequency density**. The frequency density associated with any class interval is given by:

$$\text{Frequency density} = \frac{\text{frequency in interval}}{\text{width of interval}}.$$

For example, since the first class interval 0–4 in our example is of width 5 years and has frequency 17, the frequency density for this interval is $17/5 = 3.4$ per year. Since the second class interval 5–14 is of width 10 and has frequency 43, the frequency density for this interval is $43/10 = 4.3$ per year. The frequency densities of the class intervals in Table 5.17 are shown, to two decimal places, in Table 5.18. Each frequency density has been obtained by dividing the class interval frequency by the width of the interval.

Table 5.18 *Frequency Densities for the Data in Table 5.17*

Age (years)	Frequency	Interval width (years)	Frequency density
0–4	17	5	3.40
5–14	43	10	4.30
15–29	60	15	4.00
30–44	56	15	3.73
45–64	68	20	3.40
65–79	39	15	2.60
80–99	17	20	0.85

The corresponding histogram is displayed in Figure 5.9. Here the rectangles above the class intervals have been drawn with heights proportional to the frequency densities. Observe that the area of each rectangle is numerically equal to the product of its base-length (the class interval width) and its height (in frequency density units). Also, the total area of the histogram is equal to the total frequency (300 in our example).

Figure 5.9 *Histogram for the Data in Table 5.17*

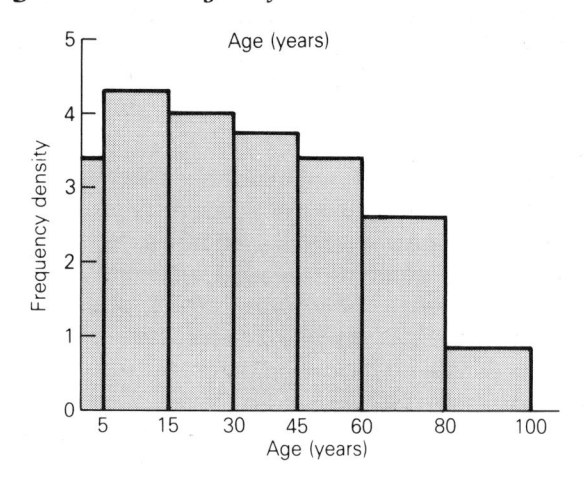

Figure 5.10 *Cumulative Frequency Polygon for the Data in Table 5.19*

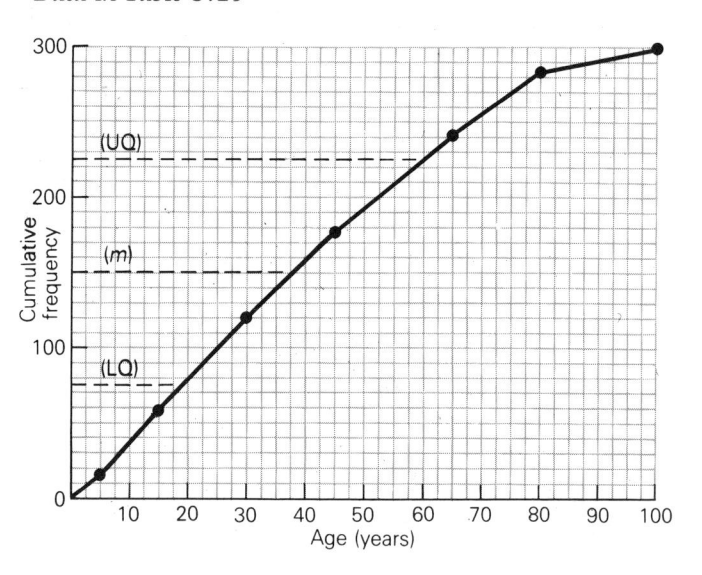

☐☐ SUMMARY MEASURES

The definitions and procedures given earlier for estimating summary measures are equally valid when the class intervals in a grouped frequency distribution are not all of the same width.

Consider Example 5.10 in which we have a grouped frequency distribution (of the ages of 300 females) with class intervals of varying widths, and we wish to estimate some percentiles of the distribution. The first step is to construct the cumulative frequency distribution (or cumulative relative frequency distribution), as we did in Section 5.4, and then to graph it. Remember that cumulative frequencies are associated with the upper end points of the class intervals. The cumulative frequency distribution for the data in Table 5.17 is shown in Table 5.19, and the corresponding cumulative frequency polygon is shown in Figure 5.10. Since the total number of observations is 300 then the median m, the lower quartile LQ, and the upper quartile UQ are estimated as the variable values for which the cumulative frequency has the values $\frac{1}{2} \times 300 = 150$, $\frac{1}{4} \times 300 = 75$, $\frac{3}{4} \times 300 = 225$ respectively. Reading from the graph in Figure 5.10 the estimates are, to the nearest year;

$$m = 37 \text{ years,}$$
$$\text{LQ} = 19 \text{ years,}$$
$$\text{UQ} = 59 \text{ years.}$$

The estimated semi-interquartile range of the distribution is:

$$\text{SIQR} = \tfrac{1}{2}(59-19) = 20 \text{ years.}$$

The cumulative frequency graph may also be used to estimate any other percentile, or how many of the 300 females are under any specified age. (See Section 5.4).

As an alternative to drawing the cumulative frequency graph we could use linear interpolation to obtain estimates of the percentiles or the number of females under a specified age. (See Section 5.5, p.56.)

Estimates of the mean and the standard deviation of the ages of the 300 females may be obtained exactly as described in Section 5.6, by replacing each class interval by its mid-value and operating on the frequency distribution of the mid-values. For the present example the relevant frequency distribution is shown in Table 5.20. Note, for example, that the age-group 0–4 corresponds to actual ages in the class interval 0–5, whose mid-value is $\tfrac{1}{2}(0+5) = 2.5$. The age-group 5–14 corresponds to actual ages in the class interval 5–15, whose midvalue is $\tfrac{1}{2}(5+15) = 10$. The other mid-values are obtained

Table 5.19 *Cumulative Frequency Distribution for the Data in Table 5.17*

Upper age limit (years)	CF
5	17
15	60
30	120
45	176
65	244
80	283
100	300

Table 5.20 *Frequency Distribution of the Mid-Values of the Age-Groups in Table 5.17*

Age-group (years)	Mid-value (x)	Frequency (f)
0–4	2.5	17
5–14	10	43
15–29	22.5	60
30–44	37.5	56
45–64	55	68
65–79	72.5	39
80–99	90	17
		Total 300

similarly. Let x denote an arbitrary mid-value and f its frequency. We find from Table 5.20 that:

$n = \Sigma f = 300$, $\Sigma fx = 12\,020$, and $\Sigma fx^2 = 661\,925$.

Hence the estimated mean age of the 300 females is:

$$\bar{x} = \frac{12\,020}{300} = 40.07 \text{ years} \quad \text{(to 2 d.p.),}$$

and the estimated standard deviation of the ages is:

$$SD = \frac{1}{n}\sqrt{(n\Sigma fx^2 - (\Sigma fx)^2)}$$
$$= \frac{\sqrt{((300 \times 661\,925) - 12\,020^2)}}{300}$$
$$= 24.52 \text{ years} \quad \text{(to 2 d.p.).}$$

☐☐ OPEN-ENDED INTERVALS

It sometimes happens that a published grouped frequency distribution includes a class interval which is open-ended at one (or both) of the extremes. For example, in a distribution of people's ages the first interval may appear as 'under 5' and/or the last interval may appear as '80 and over'.

The presence of such open-ended intervals will not matter when estimating a percentile unless it happens to be so low or so large that it falls within an open-ended interval. There should be no difficulty in estimating the median and the quartiles. For instance, the values of the median and the quartiles would remain unchanged in Example 5.9 if the first interval had been quoted as 'under 5' and the last interval as '80 and over'; the cumulative frequency polygon would be as shown in Figure 5.10, except that the point (100, 300) would not be plotted. However, there is a problem when it comes to drawing the histogram and estimating the mean and the standard deviation. These things can only be done by judiciously guessing the values of the missing class limits, and then proceeding as if there were no open-ended intervals.

● *EXERCISE 5.7*

1 Table 5.21 shows a grouped frequency distribution of the take-home pay, of 85 employees at a factory.
(a) Present the distribution as a histogram.
(b) Estimate the median and the semi-interquartile range, giving each answer to the nearest pound.
(c) Estimate the mean and the standard deviation, giving each answer to the nearest penny.

Table 5.21

Take-home pay (£)	No. of employees
80–	10
90–	24
95–	36
100–	8
105–	4
110–	3
120–	0
	Total 85

2 A grouped frequency distribution for the marks obtained by 500 candidates in an examination is shown in Table 5.22. (Treat the marks as being values of a continuous variable rounded to the nearest integers.)

Table 5.22

Mark	No. of candidates
1–20	9
21–30	52
31–40	109
41–50	122
51–60	98
61–70	83
71–80	21
81–100	6

(a) Draw a histogram of the distribution.
(b) Estimate the median and the semi-interquartile range of the marks.
(c) Estimate the 40th and the 60th percentiles of the marks.
(d) Estimate, to the nearest whole number, the percentage of the candidates that scored 65 or fewer marks.

*Projects

1 Try to get hold of a copy of the 1981 Census report on your County. If your school does not have a copy there should be one in the County public library and in the offices of your County Authority.

In Table 5 of the report you will find grouped frequency distributions of the ages of the people present in the County and in each District within the County on the night of the Census. Use the methods of this chapter to analyse and to make comparisons of the distributions within the County and in the District in which you live, and possibly in one other District within the County.

2 Obtain a recent issue of the Annual Abstract of Statistics. (Your school or county library should have a copy of the most recent issue.) Choose any one of the tables which gives a grouped frequency distribution. Analyse the distribution using the methods of this chapter.

3 Request the permission of the manager of a supermarket to observe and record the total bill for each person coming through a checkout. Having recorded the bills for at least 50 persons, compile a grouped frequency distribution of the amounts. Exhibit the distribution as a histogram and calculate statistics appropriate for summarizing the main features of the distribution.

You could also make a record of the number of items on each bill, compile a frequency distribution of the numbers, present it diagrammatically and calculate appropriate summary measures.

As an extension to this project you could consider making the observations on two or more separate occasions and commenting on any differences between the distributions on the separate occasions. Do not do this extension until you have read Chapter 6.

4 How good are people at estimating a length or a weight? As a specific example of investigating this:

(a) draw a straight line of some chosen length (20 cm for example) on a sheet of paper,
(b) find an object whose weight you know (an object weighing from 1 kg to 2 kg should be satisfactory; one possibility is to empty the contents of a packet of 1 kg or 2 kg of sugar into an unlabelled paper bag). Ask at least 50 people to estimate the length of the line you have drawn to the nearest millimetre, and to estimate the weight of the object to the nearest gram. Analyse your results, bearing in mind that the objective is to determine how good people are at estimating a length and a weight.

Review Exercises Chapter 5

☐ LEVEL 1

1 The durations of 50 telephone calls recorded to the nearest second were as follows:

342, 46, 227, 202, 349, 221, 368, 238, 275, 182,
356, 320, 185, 330, 140, 375, 225, 278, 170, 180,
321, 268, 303, 369, 282, 263, 254, 93, 302, 307,
87, 193, 146, 273, 385, 227, 310, 120, 311, 291,
342, 362, 153, 302, 318, 197, 135, 339, 251, 179.

(a) Write down the interval of the actual duration of a call that was recorded as having a duration of 46 seconds.
(b) Using a tally chart, determine the grouped frequency distribution of the actual durations of the calls, taking the class intervals as:

40.5–100.5, 100.5–160.5, 160.5–220.5, 220.5–280.5, 280.5–340.5, 340.5–400.5.

(c) Present the grouped frequency distribution as a histogram.
(d) Copy Table 5.23 and fill in the blanks to show

Table 5.23

Duration (s)	Cumulative frequency
100.5	3
160.5	
220.5	
280.5	
340.5	
400.5	50

the cumulative frequency distribution corresponding to the grouped frequency distribution in (b).
(e) Draw a graph of the cumulative frequency polygon and use it to estimate (i) the median, (ii) the semi-interquartile range, (iii) the percentage of calls that had a duration of less than 200 seconds.

2 Table 5.24 shows a grouped frequency distribution of the distances driven, in thousands of miles, by 100 car drivers during a particular year.

Table 5.24

No. of miles driven ('000)	No. of car drivers
1–	6
5–	14
9–	20
13–	36
17–	15
21–	9
25–	0

(a) Present the distribution as a histogram.
(b) Write down the mid-values of the class intervals.
(c) Denoting the mid-value of a class interval by x, and the frequency for that interval by f, you are given that $\Sigma fx = 1368$, and $\Sigma fx^2 = 21\,436$. Calculate estimates of the mean and the standard deviation of the mileages driven by the 100 car drivers, giving each answer correct to the nearest mile. Why can your answers only be estimates?

3 A time and motion study of a certain operation at a factory gave the grouped frequency distribution of the times taken by 100 workers to complete the operation shown in Table 5.25. Times are given to the nearest second.

Table 5.25

Time taken (s)	No. of workers
10–24	5
25–39	18
40–54	31
55–69	36
70–84	10

(a) Write down the class interval of the actual times taken corresponding to the tabulated interval '10–24'.
(b) State the modal class of the actual times taken.
(c) Draw up a table of the cumulative frequency distribution of the actual times taken.
(d) Graph the cumulative frequency polygon and use it to estimate (i) the median, (ii) the semi-interquartile range, (iii) the number of the 100 workers who completed the operation in less than 50 seconds.

4 The weekly earnings of 40 employees at a factory were found to be such that 5 of them earned under £80, 18 of them earned under £120, and 35 of them earned under £150. By drawing an appropriate graph, or otherwise, estimate (a) the median and the semi-interquartile range of the weekly earnings, giving each answer to the nearest pound, and (b) the number of the 40 employees who earned under £90.

5 Table 5.26 shows a grouped frequency distribution of the times, in seconds to the nearest tenth of a second, taken by 24 competitors in 3 heats of a race over 1500 metres.

Table 5.26

Time (s)	Frequency
213.1–215.0	2
215.1–217.0	7
217.1–219.0	10
219.1–221.0	3
221.1–223.0	2

(a) State the upper limits of the intervals of actual times corresponding to the intervals in the table.
(b) Draw a graph of the cumulative frequency polygon of the actual times and use it to obtain estimates of (i) the median and the semi-interquartile range of the actual times taken, (ii) the number of competitors who took less than 216 seconds, (iii) the number of competitors who took 220 seconds or longer.
(c) Only the fastest eight runners go on to the final of the race. Find the time which had to be bettered by the 8 finalists.

☐ **LEVEL 2**

6 A grouped frequency distribution of the breaking strains of 150 ropes is given in Table 5.27.
(a) Display the distribution as a histogram.
(b) Let x denote the mid-value and f the frequency

of a class interval. Applying the transformation $y=(x-125)/25$, you are given that $\Sigma fy=1589$, and $\Sigma fy^2=22837$. Deduce estimates of the mean and the standard deviation of the breaking strains of the 150 ropes, giving each answer to the nearest kilogram.

Table 5.27

Breaking strain (kg)	No. of ropes
100–	16
150–	33
300–	50
500–	41
600–	10
700–	0

7 Table 5.28 (extracted from Table 5.10 of the 1985 Edition of Social Trends) shows the values of selected percentiles of the distribution of the percentage of men's earnings taken in income tax in Great Britain in the tax year 1983/84.
(a) Write down the value of a measure of location that may be used here.
(b) Determine the value of a measure of dispersion that may be used here.
(c) By drawing an appropriate graph, or otherwise, estimate (i) the 40th and the 60th percentiles of the distribution, (ii) the percentage of men who paid less than 18% of their earnings in income tax, (iii) the percentage of men who paid between 15% and 20% of their earnings in income tax.

Table 5.28 *Percentage of Men's Earnings taken in Income Tax in 1983/1984*

Percentile	% of earnings
Lowest decile	14
Lower quartile	17
Median	20
Upper quartile	22
Highest decile	24

8 Table 5.29 (extracted from Table 5.2 of the 1985 edition of the Annual Abstract of Statistics) shows a grouped frequency distribution of the ages, in complete years, of the pupils attending schools in the United Kingdom at the beginning of January 1983.
(a) Estimate the median and the semi-interquartile range of the distribution.
(b) Assuming that the last interval is 18–20, find estimates of the mean and the standard deviation of the ages.
(Give all your answers in years and months to the nearest month.)

Table 5.29

Age (complete years)	No. of pupils ('000)
2–4	823
5–10	4302
11	879
12–14	2660
15	903
16	312
17	188
18+	27

9 A distribution of the times, in minutes, taken by 200 employees to travel to work at a factory is shown in Table 5.30.
(a) State the shape of this distribution.
(b) Estimate (i) the median, (ii) the semi-interquartile range, (iii) the 40–60 percentile range, and (iv) the number of the employees whose travel times were between 24 minutes and 48 minutes.

Table 5.30

Travel time (min)	No. of employees
under 10	5
10–	70
20–	50
30–	35
40–	25
50–	10
60 and over	5

6

COMPARING NUMERICAL DATA SETS*

"A career in profession A
is better than one in profession B
because the average salary
for those employed in A is £10,600,
compared to only £8,500 for those
employed in B."

6.1 Introduction

It frequently happens that two (or more) data sets are collected in order to compare and contrast them. Such comparisons for qualitative data were covered in Section 2.4, p. 11. Now consider data sets on a quantitative variable, which may be discrete or continuous. For example, observations of a quantitative variable may be made under two different sets of circumstances, made during two distinct periods of time, or relate to two different groups. Using the data in Table 1.2, p. 3, we might wish to investigate differences between the 20 boys and the 16 girls with respect to any one or more of the quantitative variables shoe size, family size, height and weight. (Comparing their heights is done in Example 6.2 below). In this chapter we describe methods for comparing two related data sets by means of diagrams and by means of summary measures.

6.2 Diagrammatic Comparisons

○ **EXAMPLE 6.1**

Table 6.1 shows grouped frequency distributions of the ages, in complete years, of females and males killed in road accidents in Camfordshire in 1986. (The distribution for males is the one given in Table 5.10, p. 55.)

The first observation to make from Table 6.1 is that there were more males killed (78) than females (48), so that direct comparisons of the age-group frequencies will give a misleading indication of the relative risks of females and males being killed in road accidents. As in Section 2.4 we can take account of the different totals by working with relative frequencies (or proportions). The age-group relative frequencies for females killed are obtained by dividing the frequencies by 48; those for males killed are obtained by dividing the frequencies by 78. The age-group relative frequencies, abbreviated as RF, are given to three decimal places in Table 6.2.

Table 6.1 *Grouped Frequency Distributions of the Ages of Females and Males Killed in Road Accidents in Camfordshire in 1986*

Age (years)	No. of females killed	No. of males killed
0–19	13	22
20–39	18	28
40–59	7	19
60–79	6	6
80–99	4	3
Total	48	78

Table 6.2 *Relative Frequency Distributions from Table 6.1*

Age (years)	RF$_{females}$	RF$_{males}$
0–19	0·271	0·282
20–39	0·375	0·359
40–59	0·146	0·244
60–79	0·125	0·077
80–99	0·083	0·038
Total	1·000	1·000

Comparing the last two columns of the table we can see that the two distributions have similar patterns. In particular, in each distribution the highest proportion (RF) occurs in the age-group 20–39, and the lowest proportion occurs in the age-group 80–99. The main differences are that in the age-group 40–59, the proportion of males killed (0.244) is much higher than the proportion of females killed (0.146), and that in the age-groups 60–79 and 80–99, the proportions of males killed (0.077 and 0.038 respectively) are lower than the proportions of females killed (0.125 and 0.083 respectively).

The similarities and differences between the two distributions may also be exhibited diagrammatically by means of **relative frequency histograms** as shown in Figure 6.1. It is clear from Figure 6.1 (and from Table 6.1) that both distributions are positively skew.

Figure 6.1 *Relative Frequency Histograms for the Data in Table 6.2*

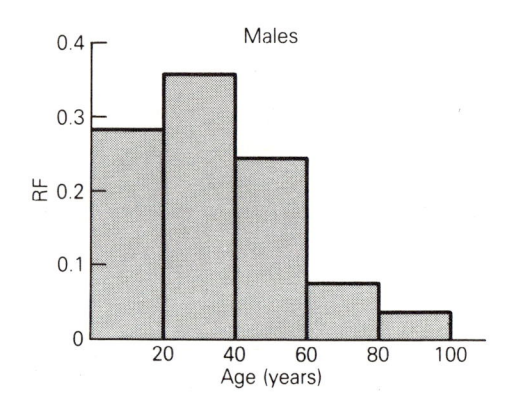

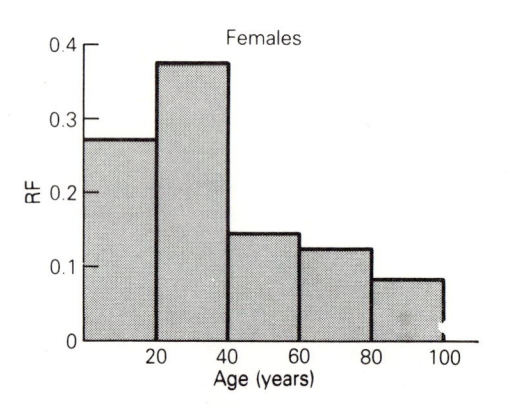

● EXERCISE 6.1

1 Table 6.3 shows the frequency distribution of the weights, recorded to the nearest 0·5 g, of the sweets contained in 60 tubes of Smarties. Construct the relative frequency distribution and compare it with that of the weights found in 36 tubes considered earlier. (Use Table 5.2, p. 49 to derive the relative frequency distribution of the weights of the sweets in the 36 tubes.) Draw relative frequency histograms and write down any differences between the two distributions that are apparent from your diagrams.

2 Table 6.4 shows grouped frequency distributions of the ages, in complete years, of males and females aged 65 and over at a certain date in one county. Draw diagrams suitable for making visual comparisons of the two distributions. Note any differences that are apparent from your diagrams.

Table 6.3

Recorded weight (g)	No. of tubes
33·5	3
34·0	9
34·5	10
35·0	13
35·5	10
36·0	9
36·5	4
37·0	2
Total	60

Table 6.4

Age last birthday	No. of males ('000)	No. of females ('000)
65–69	88	1136
70–74	577	921
75–79	367	677
80–84	192	399
85–89	74	182
90–94	18	47
Total	2028	3362

6.3 Comparing Summary Measures

For quantitative assessments of differences between numerical data sets we may compare their general levels of magnitude or 'average' values (using a measure of location) and their variabilities (using a measure of dispersion).

○ *EXAMPLE 6.2*

From Table 1.2, p. 3 the heights, to the nearest centimetre, of 20 boys and 16 girls are as follows:

Boys' heights:

159, 167, 170, 170, 174, 173, 162, 162, 171, 166, 172, 168, 165, 166, 163, 170, 172, 173, 175, 171.

Girls' heights:

152, 169, 147, 169, 150, 147, 163, 142, 145, 148, 157, 155, 160, 170, 145, 154.

Now, it is well known that for boys and girls of roughly the same ages (as in our example), the boys will tend to be taller than the girls, but what about the dispersions of their heights? We shall use our data to confirm the former and to investigate the latter.

Any measure of location is appropriate for comparing the general magnitudes of the observations in data sets. Let us first look at the medians. To this end, we first order the boys' heights and the girls' heights from shortest to tallest, to get:

Boys' heights (ordered):

159, 162, 162, 163, 165, 166, 166, 167, 168, 170, 170, 170, 171, 171, 172, 172, 173, 173, 174, 175.

Girls' heights (ordered):

142, 145, 145, 147, 147, 148, 150, 152, 154, 155, 157, 160, 163, 169, 169, 170.

Since there are 20 boys, their median height m_b is given by the average of the 10th and the 11th observations:

$$m_b = \tfrac{1}{2}(170+170) = 170 \text{ cm}.$$

Since there are 16 girls, their median height m_g is given by the average of the 8th and the 9th observations:

$$m_g = \tfrac{1}{2}(152+154) = 153 \text{ cm}.$$

We see that m_b is greater than m_g, as anticipated.

Suppose that we now decide to use the semi-interquartile ranges as measures of dispersion in conjunction with the medians. (We could equally well have chosen the mean absolute deviations from the median.) We can find the lower and upper quartiles by applying the rules given on p. 24. For the boys $\tfrac{1}{4} \times 20 = 5$, so $k=5$ and the lower and upper quartiles of the boys' heights can be found as follows. LQ_b is the average of the 5th and the 6th observations:

$$LQ_b = \tfrac{1}{2}(165+166) = 165.5 \text{ cm}.$$

UQ_b is the average of the 15th and the 16th observations:

$$UQ_b = \tfrac{1}{2}(172+172) = 172 \text{ cm}.$$

Thus the semi-interquartile range of the boys' heights is given by:

$$SIQR_b = \tfrac{1}{2}(172-165.5) = 3.25 \text{ cm}.$$

Similarly, for the girls, we have $\frac{1}{4} \times 16 = 4$, so $k=4$ and the lower and upper quartiles of the girls' heights can be found as follows. LQ_g is the average of the 4th and the 5th observations:

$$LQ_g = \tfrac{1}{2}(147 + 147) = 147 \text{ cm}.$$

UQ, is the average of the 12th and 13th observations:

$$UQ_g = \tfrac{1}{2}(160 + 163) = 161.5 \text{ cm}.$$

It follows that the semi-interquartile range of the girls' heights is given by:

$$SIQR_g = \tfrac{1}{2}(161.5 - 147) = 7.25 \text{ cm}.$$

Since $SIQR_g$ is more than twice $SIQR_b$, the girls' heights are much more variable (dispersed) than the boys' heights.

We can carry out similar comparisons using the means and the standard deviations. Letting b denote

TOTAL CONFUSION

THE MEDIA often emphasise total numbers of deaths in traffic fatality statistics, particularly over holiday periods. This gives the impression that travel has become steadily more dangerous over the years.

In fact, the opposite is true as can be seen from the continuously decreasing death rate per vehicle or per passenger mile.

Overemphasizing totals in this way may lead to inappropriate corrective measures. For example, laws to reduce speed limits, taking no account of increased traffic, may result in greater traffic hazards. It may, for example, cause greater traffic jams, thereby delaying people trying to reach hospital. This could turn 'serious injury' statistics into 'fatality' statistics!

Totals are sometimes used to make a change look large. To minimize a change, rates per some small unit may be used. For example, in a period of 5% inflation, a union may ask for a 7% rise. If the average hourly wage is £4, they will describe this as a 28 pence rise. Reducing anything to pence rather than pounds creates an impression of smallness.

The company, on the other hand, will offer a 6% rise, pointing out that for its 10 000 employees, working 2000 hours per year, this comes to an increased cost of £4·8 million. The innocent bystander who does not keep an eye on the comparisons' bases may find it difficult to keep track of £4·8 million being less than 28 pence. Changing, or unmentioned, bases tend to confuse the receiver of such quantitative arguments.

a boy's height we find that:

$$\Sigma b = 3369 \quad \text{and} \quad \Sigma b^2 = 567\,897,$$

so that the mean and the standard deviation of the boys' heights are:

$$\bar{b} = \frac{3369}{20} = 168.45 \text{ cm},$$

and:

$$SD_b = \frac{\sqrt{((20 \times 567\,897) - 3369^2)}}{20} = 4.41 \text{ cm}$$

(to 2 d.p.).

Letting g denote a girl's height we find that:

$$\Sigma g = 2473 \quad \text{and} \quad \Sigma g^2 = 383\,521,$$

so that the mean and the standard deviation of the girls' heights are, to 2 decimal places:

$$\bar{g} = \frac{2473}{16} = 154.56,$$

and

$$SD_g = \frac{\sqrt{((16 \times 383\,521) - 2473^2)}}{16} = 8.97 \text{ cm}.$$

Our conclusions are exactly as before, namely that the boys tend to be taller than the girls and the boys' heights are much less dispersed than the girls' heights.

○ EXAMPLE 6.3

Compare the locations and the dispersions of the two distributions in Example 6.1.

Let us choose the means and the standard deviations for the required comparisons. (You are asked to consider the medians and semi-interquartile ranges in Question 1, Exercise 6.2.)

We saw in Section 5.6, p. 58 that estimates of the mean and the standard deviation of a grouped frequency distribution can be obtained from the corresponding values of the frequency distribution of the mid-values of the class intervals. From Table 6.1, the frequency distributions of the mid-values of the class intervals are as displayed in Table 6.5.

In Example 5.8, p. 60 we showed that the estimated mean and standard deviation of the ages of the males killed were to 2 decimal places:

$$\bar{x} = 34.62 \text{ years},$$

and:

$$SD(x) = 21.22 \text{ years}.$$

Denoting a female's age by y we find that:

$$\Sigma fy = 1800 \quad \text{and} \quad \Sigma fy^2 = 96\,800.$$

Table 6.5 *Frequency Distributions of the Mid-Interval Values from Table 6.1*

Mid-interval age (years)	No. of females killed	No. of males killed
10	13	22
30	18	28
50	7	19
70	6	6
90	4	3
Total	48	78

Hence the mean and the standard deviation of the females' ages are:

$$\bar{y} = \frac{1800}{48} = 37.5 \text{ years},$$

and:

$$SD(y) = \frac{\sqrt{(48 \times 96\,800 - 1800^2)}}{48} = 24.71 \text{ years}$$

(to 2 d.p.).

Comparing the means and the standard deviations of the two distributions we see that the average and the dispersion of the female's ages are somewhat higher than those of the males' ages. Quantitative comparisons of this nature are not possible from the relative frequency histograms.

○ EXERCISE 6.2

1 Compare the medians and the semi-interquartile ranges of the two distributions in Table 6.1, p. 68.

2 Compare the means and the standard deviations of the weights of the 20 boys and the 16 girls given in Table 1.2, p. 3. (Keep your answers for use in Question 2, Exercise 6.3).

Table 6.6

Age at marriage (years)	No. of males ('000)	No. of females ('000)
under 21	31	84
21–24	111	116
25–29	88	62
30–34	41	29
35–44	40	30
45–54	17	13
55+	15	10

3 *Table 6.6, extracted from Table 2.1 of the 1985 edition of the Annual* Abstract of Statistics, *shows the distributions of the ages at marriage of all married males and females in England in 1983. Summarize the two distributions by calculating appropriate summary measures and note any differences between them.*

6.4 The Coefficient of Variation

One shortcoming of a measure of dispersion is that its magnitude will depend upon the unit of measurement of the variable. In Example 6.2 we considered the heights of 20 boys measured in centimetres and found, for example, that the standard deviation of the heights was approximately 4.4 cm. Now if instead the heights had been measured in metres, the standard deviation would be 0.044 m. This illustrates that the larger the unit of measurement the smaller will be the magnitude of the standard deviation. This is also true of any other measure of dispersion.

It would be preferable to have a measure of dispersion whose value is independent of the unit of measurement. The most widely used such measure is the **coefficient of variation**, which expresses the standard deviation of a data set as a percentage of its mean. Thus the coefficient of variation (CV) of a data set is defined as:

$$CV = \frac{\text{standard deviation}}{\text{mean}} \times 100.$$

(Similar quantities for use with other measures of location and dispersion have also been proposed but are rarely used in practice.)

Observe that CV is dimensionless, that is, it has no units. This means that its value does not depend upon the unit of measurement. Consider the boys' heights in Example 6.2. The mean and the standard deviation of the heights have been shown to be 168.45 cm and 4.41 cm respectively; thus the coefficient of variation is:

$$CV = \frac{4.41}{168.45} \times 100 = 2.62\% \quad \text{(to 2 d.p.)}.$$

Had the heights been measured in metres, the coefficient of variation would be:

$$CV = \frac{0.0441}{1.6845} \times 100 = 2.62\% \quad \text{(to 2 d.p.)},$$

exactly as before, thus illustrating the independence of the value of CV of the unit of measurement.

However, the value of CV does depend on the choice of origin (because of the mean), and consequently is not appropriate for a data set of observations of a variable measured on an interval scale. (See Section 1.4 p. 5.) For example, the value of the CV of a data set on temperatures will be different for temperatures measured on the Celsius scale and the Fahrenheit scale.

As another example to demonstrate the need for a quantity such as the coefficient of variation, consider a machine which is set to automatically deliver 10 cc of a liquid into phials. What happens in practice is that the amounts of the liquid delivered to the phials vary slightly around 10 cc. Suppose that the standard deviation of the amounts delivered is equal to 1 cc, which may be regarded as a measure of the precision of the machine. Now consider another machine which is set to automatically deliver 1000 cc of liquid into bottles and that the standard deviation of the amounts delivered is also equal to 1 cc. Although the standard deviations (dispersions) of the amounts delivered are the same for both machines, the second machine is much more precise than the first machine because 1 cc is a much smaller quantity relative to 1000 cc than to 10 cc. This difference may be quantified by means of the coefficients of variation. For the two machines under consideration the coefficients of variation (or measures of their precisions in this case) are as follows. For the first machine:

$$CV_1 = \frac{1}{10} \times 100 = 10\%,$$

For the second machine:

$$CV_2 = \frac{1}{1000} \times 100 = 0.1\%.$$

Note that the smaller the value of the coefficient of variation the greater is the precision as considered here.

The following examples illustrate another situation in which the coefficient of variation is an appropriate quantity for comparing the relative variations of two data sets having very different means.

○ *EXAMPLE 6.4*

The annual salaries of a sample of accountants in the U.K. had a mean of £12 565 and a standard deviation of £2410, while the salaries of a sample of accountants in the U.S.A. had a mean of $23 500 and a standard deviation of $2060. Which sample shows the greater relative variability?

The coefficients of variation for the two samples are, to two decimal places:

$$CV_{U.K} = \frac{2410}{12\,565} \times 100 = 19.18\%$$

$$CV_{U.S.A.} = \frac{2060}{23\,500} \times 100 = 8.77\%$$

Although the standard deviations (£2410 and $2060) are of the same order of magnitude, it is clear from the values of the coefficients of variation that the relative variation of the salaries of the U.K. accountants is more than double that of the U.S.A. accountants. (To compare the averages and the standard deviations we would need to know the current exchange rate between dollars and pounds, but such comparisons could be misleading without taking into account the relative costs of living in the U.K. and the U.S.A.)

○ *EXAMPLE 6.5*

For the heights of the boys and girls given in Example 6.2 which data set has the greater relative variability?

From the results obtained in Example 6.2 the coefficients of variation of the boys' heights and the girls' heights are, to two decimal places:

$$CV_{boys} = \frac{SD_b}{\bar{b}} \times 100 = \frac{4.41}{168.45} \times 100 = 2.62\%,$$

and

$$CV_{girls} = \frac{SD_g}{\bar{g}} \times 100 = \frac{8.97}{154.56} \times 100 = 5.80\%$$

The ratio of these coefficients of variation does not differ much from the ratio of the standard deviations; this is because the means of the data sets are not very different. Consideration of relative variations is only relevant when comparing data sets having very different means. This was the case in Example 6.4 where the ratio of the coefficients of variation was very different from the ratio of the standard deviations.

● *EXERCISE 6.3*

1 A sample of Brand A electric light bulbs had lifetimes whose mean was 1205 hours and whose standard deviation was 280 hours. A sample of brand B 'double life' electric light bulbs had lifetimes whose mean and standard deviation were 2095 hours and 320 hours respectively. Which sample shows (a) the greater dispersion in their lifetimes, (b) the greater relative dispersion in their lifetimes?

2 Compare the relative dispersions of the weights of the 20 boys and the 16 girls given in Table 1.2, p. 3. (Use the results you obtained in answer to Question 2, Exercise 6.2.)

3 The mileages travelled by 80 car drivers during the past year had the grouped frequency distribution shown in Table 6.7. Do these drivers show greater (or less) relative variation in the mileages they have driven during the past year as compared with the numbers of years they have been driving whose mean is 16 and standard deviation is 4?

Table 6.7

Miles travelled	No. of drivers
3–	4
6–	10
9–	14
12–	31
15–	12
18–	6
21–	3
24–	0

4 In 1970 the prices of new houses in a certain locality averaged £7500 with a standard deviation of £2400, while in 1985 in the same locality the prices averaged £33 800 with a standard deviation of £6900. Determine in which of the two years the prices of new houses had the larger relative dispersion.

Review Exercises Chapter 6

1 A firm has 100 machines of type A and 40 machines of type B. Table 6.8 shows, for each type of machine, a grouped frequency distribution

Table 6.8

Time to breakdown (hours)	No. of type A machines	No. of type B machines
0–	4	2
50–	18	9
100–	60	17
150–	10	10
200–	5	2
250–	2	0
300–	1	0
350–	0	0

of the operational times, in hours, until the first breakdown.

(a) Draw histograms appropriate for a visual comparison of the two distributions. Note any differences between the distributions.

(b) On the same sheet of graph paper and using the same axes and scales, draw the cumulative relative frequency polygons of the distributions. Using your graph, or otherwise, estimate the medians and the semi-interquartile ranges of the distributions. Interpret the differences between these values for the distributions.

(c) Using your graphs, estimate (i) the two percentiles which have the same values for both distributions and state their common value, (ii) the difference between the 60th percentiles of the two distributions.

2 Table 6.9 shows a grouped frequency distribution of the amounts spent on food in a year by 100 families.

(a) Estimate the mean and the standard deviation of the amounts spent per family.

(b) If the amounts spent by the same families on clothing had mean £900 and standard deviation £300, compare the relative variations in the amounts spent on food and on clothing.

3 A micrometer was used repeatedly to measure the diameter of a ball bearing and it was found that the measurements had mean 5.73 mm and standard deviation 0.015 mm. Another micrometer was used repeatedly to measure the diameter of a metal ball and it was found that these measurements had mean 8.56 cm and standard deviation 0·04 cm. Determine which of the micrometers has the greater precision of measurement.

4 Table 6.10 shows the frequency distributions of the number of faulty tyres per car in spot checks on 26 cars that were under 5 years old and 34 cars that were 5 or more years old.

(a) Draw diagrams suitable for a visual comparison of the two distributions.

(b) Calculate the mean, the standard deviation, and the coefficient of variation for each distribution. Using the values you obtain, comment on any differences there are between the two distributions.

Table 6.9

Amount spent on food (£)	No. of families
1500–	6
1900–	15
2300–	22
2700–	34
3100–	15
3500–	6
3900–	2
4300–	0

Table 6.10

No. of faulty tyres	No. of cars under 5 years old	No. of cars 5 or more years old
0	17	5
1	5	15
2	3	9
3	0	4
4	1	0
5	0	1

7
POPULATIONS, SAMPLES, AND QUESTIONNAIRES

"Are you a lady
or a gentleman?"

*Bar Hill Community Primary School,
1986 Annual Applied Statistics entry.*

7.1 Populations and Samples

A statistical investigation is one designed to obtain information on some characteristics (variables) of a collection of individuals. The collection of interest is referred to as the **target population**, or simply as the **population**. The most familiar population is the collection of people living in a region (a village, town, city, county, or country). In Example 1.2, p. 2 the collection of pupils in a particular class was the population of interest. In Statistics the population need not be a collection of persons. It may be a collection of plants, animals, manufactured items, farms, supermarkets, amusement arcades, and so on.

When the required information is to be obtained from *all* the individuals in a population, the investigation is called a **census**. Despite the magnitude of the task, the government holds a census of all the residents of the United Kingdom every ten years, the last one having been in 1981.

Generally, when the target population is large a census is ruled out as being too time-consuming or too costly or both. Instead the required information is only obtained from some of the individuals of the population. The chosen individuals are referred to as a **sample** from the population, and the investigation itself is referred to as a **sample survey**. In order to speed up the processing of the data obtained in the 1981 U.K. census only a sample of 10% of the returns were used on some of the variables (for example, occupation, workplace, higher qualifications).

Sample surveys are very common these days. They are used by local government to obtain information needed for future planning. They are also used in market research, for example, to obtain information on the numbers of viewers of various television programmes, on people's reactions to some new product, and on voters' strength of support for the various political parties or for some government proposal.

In Example 1.1, p. 1 we only had a sample of tubes of Smarties available. Here we have an instance where it would not be possible to carry out a census, the population being *all* the tubes produced. In the discussion of the results obtained from our sample we mentioned that the results should not be extended to apply to all tubes unless we were confident that our sample was *representative* of all tubes produced. It will always be the case that generalization from a sample to the population will be valid only when the sample is representative of the population in the characteristics examined. One example where this is so is a blood sample taken from a person for testing; it is assumed that the sample is typical of that person's blood. However, it is only rarely that a sample is known to be representative of the population.

Unfortunately there is no known method of selecting a sample which is guaranteed to be representative of a population. So, how should a sample be chosen? In the first place it is clear that we should avoid **selection bias**. This arises when some particular individuals in the population are favoured more than others when making the selection. A sample chosen in this way is most unlikely to be representative. It is well known that a selection procedure based on human judgement is likely to be biased, even though the person making the selection may not introduce the bias consciously.

The safest way of eliminating selection bias is to have some mechanical procedure which ensures that all the individuals in the population have the same chance of being selected. Such a sample is also likely (but not guaranteed) to be representative of the population. Such a procedure is referred to as **random sampling** and the chosen individuals are called a **random sample** from the population. Random sampling is described in the next section.

● **EXERCISE 7.1**

1 *What is the difference between a census and a sample survey?*
2 *What is meant by the target population?*
3 *What is the advantage in having a sample that is representative of the population?*
4 *A teacher requires a sample of 5 pupils from her class for some specific purpose. She chooses the 5 sitting in the front row. Give an example of a variable that may be associated with pupils for which the selected sample would certainly not be representative of the entire class.*
5 *What is meant by selection bias? Name a safe procedure for eliminating it.*
6 *A sample survey is to be made of the reading habits of non-employed housewives in a village. The sample is to be chosen from women visiting the local library, and by obtaining the information required from those of them who are non-employed housewives. Comment on this sampling method.*

7.2 Random Samples

As mentioned in the previous section a **random sampling procedure** is one which ensures that every individual in a population has the same chance of being selected. The selected individuals are then referred to as a **random sample** from the population.

To illustrate how a random sample may be obtained, suppose a sample of 30 individuals is required from a population of 100 individuals, numbered from 1 to 100 on a list. Having prepared a pack of 100 cards numbered from 1 to 100, shuffle the pack well and deal out 30 cards. The numbers on the cards dealt are taken to be the numbers of the listed individuals that will be sampled. This gives a random sample.

The snag with this method is that it is very cumbersome when the population size is large. A more practical method is one that uses **random digits**. Random digits are a sequence of the ten digits 0, 1, . . . , 9, in which each recorded digit in the sequence is a random selection. Tables of random digits have been published, and many calculators and computers have programs for generating random digits. These digits are not strictly chosen at random since they are usually obtained from mathematical formulae. For this reason digits obtained in this way are often called **pseudo-random**. Fortunately they have been found to be as satisfactory as random digits for practical purposes when choosing a random sample.

Table 7.1 consists of 500 random digits arranged in 25 rows and 20 columns. The following example illustrates how such a table can be used.

○ **EXAMPLE 7.1**

Use Table 7.1 to choose a random sample of 6 girls from a list of 22 girls.

The first step is decide upon a starting point in the table. This is done by:

(1) arbitrarily choosing a number from 1 to 25, inclusive, to fix the starting row,
(2) arbitrarily choosing a number from 1 to 20, inclusive, to fix the starting column.

Table 7.1 *Random Digits*

Row No.	1	2	3	4	5	6	7	8	9	10	11	12	13	14	15	16	17	18	19	20
1	5	9	9	6	0	1	3	6	8	8	7	7	9	0	4	5	5	9	6	4
2	7	2	0	8	5	9	4	4	6	7	9	8	5	6	6	5	1	4	9	6
3	1	0	9	1	4	6	9	6	8	6	1	9	8	3	5	2	4	7	5	3
4	6	5	0	0	5	1	9	3	5	1	3	0	8	0	0	5	1	9	2	9
5	5	6	2	3	2	7	1	9	0	3	7	3	5	2	9	3	7	0	5	0
6	4	8	2	1	4	7	7	4	6	3	1	7	2	7	2	7	5	1	2	6
7	3	5	9	6	2	9	0	0	4	5	8	4	9	0	9	0	6	5	7	7
8	6	3	9	9	2	5	6	9	0	2	0	9	0	4	0	3	3	5	7	8
9	1	9	7	9	9	5	0	7	2	1	0	2	8	4	4	8	5	1	9	7
10	2	8	5	5	5	3	0	9	4	8	8	6	2	8	3	0	0	2	3	5
11	7	1	3	0	3	2	0	6	4	7	9	3	7	4	2	1	8	6	3	3
12	4	1	9	4	5	4	0	6	5	7	4	8	2	8	0	1	8	3	8	4
13	0	9	1	1	2	1	9	1	7	3	9	7	2	8	4	4	7	4	0	6
14	2	2	3	0	9	5	6	9	7	2	3	8	5	8	2	2	1	4	7	9
15	2	4	3	2	1	2	3	8	4	2	3	3	5	6	9	0	9	2	5	7
16	8	9	1	7	9	5	8	8	2	9	0	2	3	9	5	6	0	3	4	6
17	9	7	7	4	0	6	5	6	1	7	1	4	2	3	9	8	6	1	6	7
18	7	0	5	2	8	5	0	1	5	0	0	1	8	4	0	2	7	8	4	3
19	1	0	6	2	9	8	1	9	4	1	1	8	8	3	9	9	4	7	9	9
20	4	6	4	0	6	6	4	4	5	2	9	1	3	6	7	4	4	3	5	3
21	3	0	8	2	1	3	5	4	0	0	7	8	4	5	6	3	9	8	3	5
22	5	5	0	3	3	6	6	7	6	8	4	9	0	8	9	6	2	1	4	4
23	2	5	2	7	9	9	4	1	2	8	0	7	4	1	0	8	3	4	6	6
24	1	9	4	2	7	4	3	9	9	1	4	1	9	6	5	3	7	8	7	2
25	3	7	5	6	0	8	1	8	0	9	7	7	5	3	8	4	4	6	4	7

The entry in the chosen row and column will be the first choice of digit. Suppose we choose the 13th row and the 5th column. The entry there is 2, which is the first random digit we will use.

The next step is to arbitrarily decide on a direction in which to move from the starting point. The choices open to us are to move along the 13th row, to the left or to the right, or to move along the 5th column, upwards or downwards. Suppose our choice is to move from left to right along the 13th row. Doing so the digits we get from the table are:

$$2, 1, 9, 1, 7, 3, 9, 7,$$
$$2, 8, 4, 4, 7, 4, 0, 6.$$

This takes us to the end of the 13th row. We now continue onto the 14th row. We continue in this way until we have enough random digits for our purpose.

Since our population size, 22, has two digits we need to take the random digits in successive pairs. (For a population size of three digits the random digits would be taken in successive triples). We will stop as soon as we have got enough numbers for the required sample size, that is 6 **different** numbers from 01 to 22, inclusive.

The first pair of digits we have from the table is 21, so that the 21st girl on the list is our first choice to be included in the sample. The next pair of digits is 91. Since this is outside the range from 1 to 22 it has to be discarded. The next pair is 73, which again is outside the range. Continuing in this way the successive pairs are 97, 28, 44, 74, and 06. All but the last of these have to be discarded. Since the last pair, 06, is within the range we want, we take the 6th girl on the list as our second sampled girl. We now continue reading off successive pairs of digits from the table until we end up with 6 different numbers from 1 to 22. Doing so, the complete set of pairs of random digits obtained from the table is as follows:

21, 91, 73, 97, 28, 44, 74, **06**,
(13th row entries)
22, 30, 95, 69, 72, 38, 58, 22, **14**, 79,
(14th row entries)
24, 32, **12**, 38, 42, 33, 56, 90, 92, 57,
(15th row entries)
89, **17**.
(16th row entries)

Here the different numbers from 01 to 22 have been

[77

printed bold. Every other number has to be discarded, either because it is outside the range or because it has appeared earlier (22 appears twice in the entries from the 14th row). It follows that our sample of girls will be those numbered 21, 6, 22, 14, 12, and 17 on the list.

It can be shown that 74613 different samples of 6 girls can be chosen from 22 girls. The procedure used above ensures that every one of these possible samples has the same chance of being the chosen sample. The chance that we will end up with an unrepresentative sample (for example, the 6 tallest girls) is extremely small. Even if we are unlucky enough to end up with an unrepresentative sample, at least we will know that it has arisen completely by chance and not from any bias.

● *EXERCISE 7.2*

1 Use the method of Example 7.1 to obtain:
(a) a random sample of 8 individuals from a list of 30 individuals,
(b) a random sample of 10 individuals from a list of 80 individuals.

☐☐ THE REMAINDER METHOD

The procedure used in Example 7.1 is not very efficient. We had to generate 30 pairs of random digits in order to get 6 different numbers from 1 to 22. An alternative and quicker procedure is the **remainder method**.

In this method each pair of random digits is divided by 22 (the population size) and replaced by the remainder obtained. Any remainder obtained will be one of the numbers from 0 to 21. The remainder is then taken to be the number on the list of the girl to be included in the sample. The only exception to this is that a remainder of 0 is taken to be the *last* (22nd) girl on the list.

There is one other problem to overcome before applying this method. The pairs of digits we obtain from the table of random digits will range from 00 to 99. To ensure we have a random selection from these it is necessary that all the possible remainders occur equally often. Now for the numbers from 00 to 87, each possible remainder (00 to 21) occurs four times (once in each of 00 to 21, 22 to 43, 44 to 65, and 66 to 87). But for the other numbers, 88 to 99, the only possible remainders are from 0 to 11. So if all the pairs from 00 to 99 are included, remainders from 0 to 11 occur five times, but those from 12 to 21 occur only four times. To correct for this we discard every pair of digits from 88 upwards.

We are now ready to apply the method. We shall use the same starting point and direction of movement as before. Reading from the table of random digits and dividing each pair of digits by 22, the remainders we obtain are as follows:

21, D, **7**, D, **6**, **0**, **8**, 6,   (13th row entries)
0, 8, D, **3**.   (14th row entries)

Here, D denotes a discarded number (88 or more) and the 6 different remainders have been printed bold. (Observe also that we exclude the 6, 0 and the 8 when they occur for the second time.) It follows that the numbers of the girls in our sample will be 21, 7, 6, 22, 8, and 3. (Recall that a remainder of 0 is taken to be the last number on the list, 22 in our example.) The remainder method has given us a sample of size 6 from only 12 pairs of digits as compared with 30 pairs using the first method.

For a population of size n, where n is a two digit number, the procedure is as follows. First calculate the integer part of $100/n$. For example, if $n=16$ then $100/16 = 6\frac{1}{4}$, the integer part of which is 6; if $n=20$ then $100/20 = 5$, the integer part of which is 5; if $n=35$ then $100/35 = 2.85$ (to 2 decimal places), the integer part of which is 2.

Let k be the integer part obtained. The rule is to discard every number (pair of digits) from $k \times n$ upwards. For example, if $n=16$ then $k=6$ and $kn = 6 \times 16 = 96$, so we discard all numbers from 96 upwards (to 99); if $n=20$ then $k=5$ and $kn = 5 \times 20 = 100$, so that no number is discarded; if $n=35$ then $k=2$ and $kn = 2 \times 35 = 70$, so we discard all numbers from 70 upwards. In the above example we had $n=22$ so that $k=4$, and we discarded all numbers from $4 \times 22 = 88$ upwards.

For a population size n, where n is a three digit number (100 to 999) the random digits have to be taken in successive triples. The only change to the above rule is to replace 100 by 1000. For example, if $n=550$ then $1000/550 = 1.81$ (to 2 decimal places), so that $k=1$, and we discard every number (now of three digits) from $1 \times 550 = 550$ upwards.

2 Use the remainder method to answer (a) and (b) of Question 1.
3 Use the remainder method to obtain a random sample of size 10 from a population of size (a) 120, (b) 680.

7.3 Some Other Random Sampling Procedures

Sometimes a particular type of random sampling is more appropriate when carrying out a sample survey. We describe three such procedures in this section.

☐☐ STRATIFIED RANDOM SAMPLING

Suppose that a sample survey is to be carried out to investigate the sporting activities of the pupils in a coeducational school. Taking a random sample of pupils from the school could result in the chosen pupils all being boys (or a vast majority being boys). Since the sporting activities of boys and girls are likely to be different (more boys than girls play soccer, while more girls than boys play netball) the sample responses may not give a true picture of the sporting activities of the pupils in the school. Clearly it would be better to ensure that the sample includes reasonable numbers of both girls and boys. This can be done by choosing separate random samples of girls and boys, and combining them into one sample. Such a sample is called a **stratified random sample**. The population of all pupils has been stratified (subdivided) into two subpopulations, one of girls and the other of boys. The two subpopulations are called the **strata** of the population. In our example the variable used for stratification was sex and is referred to as the **stratifying variable**.

If the numbers of girls and boys in the school are about equal then it would be reasonable to take random samples of the same size. Otherwise the sample sizes should be proportional to the strata sizes. For example, if there are 850 pupils, of whom 580 are girls and 270 are boys, then the proportion of the pupils who are girls is 580/850 and the proportion who are boys is 270/850. If the total sample size is to be 100, say, then the size of the sample of girls should be $(580/850) \times 100 = 68$, and the size of the sample of boys should be $(270/850) \times 100 = 32$.

Stratified sampling is always preferable when it is known (or suspected) that the characteristic being investigated (sporting activities in the above example) may vary from one stratum to another. If we were to carry out a sample survey of the attitudes of employees in a factory to their working environment then we would expect the attitudes to be different for manual workers, skilled workers and professionals. This suggests that we should stratify the population of all employees into three strata; one of manual workers, one of skilled workers and one of professionals.

An important advantage of stratified sampling is that it is more likely than random sampling to give a sample that is representative of the population. It also has the advantage that comparisons can be made of the responses from the various strata. The main problem with this method is that of being able to identify appropriate stratifying variables, which should be those that are likely to have a bearing on the characteristic being investigated. In other words the strata should be such that responses from *within* a stratum are more similar than responses from different strata.

● EXERCISE 7.3

1 A sample survey is to be carried out to investigate the attitudes of the pupils at a school towards scenes of violence shown on television. It is thought that the pupils' attitudes may vary with age, so it is decided to stratify the school population according to the age-groups 11–13, 14–16, and 17+. The numbers of pupils in the school in these age-groups are 352, 284, and 103 respectively. If the total sample size is to be 120, find the sizes of the random samples to be taken from the three strata.

2 In a sample survey on the wages and salaries paid to employees at a large factory suggest two variables that may be appropriate for stratifying the population of all employees.

☐☐ CLUSTER SAMPLING

Sometimes there is a natural subgrouping of the target population. The subgroups are called **clusters**. If all members of the target population are voters then the parliamentary constituencies are natural clusters, and there will be a list of all voters in each constituency. As another example, if the target population consists of all children in the country attending primary school then the local education authorities form natural clusters; each authority should have a list of all the primary school children in its catchment area.

When a sample survey is carried out on a population which can be subdivided into clusters it is often more convenient to first choose a random sample of clusters and then to sample within each cluster chosen. Whereas strata are chosen so that responses within a stratum are expected to be more similar than responses from different strata, clusters are such that responses within a cluster are expected to be more variable than responses from different clusters.

An advantage of cluster sampling is that it can be much less costly than random sampling. This is particularly so when the clusters are geographical regions and interviewers are used. An interviewer's sample would be chosen from just one cluster. Visiting individuals within a cluster will involve much less travelling for an interviewer than visiting individuals in different clusters.

3 A sample survey is to be carried out on some characteristics of the population of boys in a school. Suggest how the population may be subdivided into clusters. State the advantage here of using cluster sampling as compared with complete random sampling.

4 State briefly the difference between strata and clusters.

☐☐ SYSTEMATIC SAMPLING

Random sampling from a very large population is cumbersome. A possible alternative is **systematic sampling**. The procedure is best illustrated by means of an example. Suppose that a sample of size 250 is required from a population of size 60 000. Since 60 000/250 = 240 the sample is to consist of 1 in every 240 of the entire population. Now choose a number at random from 1 to 240 inclusive (for example, by using a table of random digits). This will be the number on the list of the first individual to be included in the sample. The other 239 individuals to be included in the sample are then taken to be those corresponding to the numbers obtained by successively adding 240 to the chosen random number. If the randomly chosen number from 1 to 240 is 36, then the sample will consist of the individuals corresponding to the numbers 36, 36 + 240 = 276, 276 + 240 = 516 and so on, the final number being 59 796.

A systematic sample is likely to be representative of the population provided that the order of the individuals on the population list does not relate to the characteristic of interest. Suppose a sample survey is to be made of a shop's daily turnover during a year, and that the sample is to be of size 52. Since there are 6 working days per week and 52 weeks in a year the population consists of the turnovers on 6 × 52 = 312 days. Our sample is equivalent to choosing 1 in every 6 of these. For a systematic sample, a number is first chosen at random from 1 to 6; this gives one of the 6 daily turnovers in the first week for inclusion in the sample. Then adding 6 successively will give the other days to be included. Suppose that the chosen random number is 2. Then the daily turnover on the Tuesday of the first week is included in the sample. But every other chosen day will also be a Tuesday. So the sample of daily turnovers will be those on Tuesdays. This is clearly an unrepresentative sample. The problem here could be overcome by first randomly ordering the 312 days before taking a systematic sample.

Systematic sampling is very convenient in quality control. To check on the quality of manufactured items it would be reasonable to take, say, every tenth item produced in a day as a sample of items for that day. It would also be convenient for sampling from a long queue of shoppers at a checkout or at a cinema.

5 Describe how you would take a systematic sample of size 20 from a population of size 100.

6 A systematic sample of size 40 is to be drawn from a population of size 570. Explain how this may be done. What is unsatisfactory about systematic sampling in this case?

7.4 The Sampling Frame

Each of the sampling methods described so far requires the population of individuals to be on a numbered list. This is not always the case. For example, for a sample survey of voters in a constituency the electoral list will have been prepared some time prior to the last election. As a result it may well be out of date. If follows that the list is not of the target population. But it may be the best available list for the purpose of the survey. In such a situation the list that is available is referred to as the **study population** or the **sampling frame**. When the sampling is from the actual target population then the sampling frame is the target population itself.

A sampling frame other than the target population itself will be necessary whenever there is no list of the target population. Other examples in which a list of the target population may not be available include the population of all the people in a district who are in full-time employment, and the population of all the housewives in a district. In some cases there may be a list but some of the individuals may not be available at the time of the survey. For example, some people may be abroad or in hospital at the time; in a school survey some pupils may be absent for various reasons.

CENSUS DOOMSDAY

In 1987, the first West German national census for 17 years was conducted in the face of widespread opposition led by the Green Party. In places, opposition was so violent that many of the 500,000 distributors of census forms had to be accompanied on their rounds by protective police guards.

To most people, the questions on the census forms seem inoffensive. The census asked about people's age, their main source of income, details of employment and numbers of hours worked each week, how far they travel to work and by what means, what their housing is like and rend paid. Other personal questions include religion, nationality and education. The authorities claimed that the information was needed for future planning of housing, road-building, transport, energy needs, and health care and welfare provision.

In West Germany it is compulsory to carry computer-readable identity cards and to register all residents. Opponents of the census therefore claimed that the government already has sufficient data for planning welfare provisions.

Opponents also maintained that there was insufficient anonymity in the census returns. The government promised that identities would be deleted from the records shortly after the census is completed, but not all German citizens trust the authorities. Already, they complain, computerised networks run by the police distribute data on peace marchers throughout the country. The federal commissioner for data protection in West Germany says that the laws encourage mistrust, "Citizens have the right to ask to see the data that are held on them, but no one has to tell them who and where to ask."

West German statisticians claim that if the non-response rate exceeds 5%, then the census will be useless for forecasting future requirements reliably. One forecast that may be reliable, however, is that this will be the last census to be held in West Germany.

The responses obtained when the sample is from a sampling frame might still be representative of the target population, but some check will need to be made that individuals in the population who are not in the sampling frame are not markedly different from those in the frame.

The individuals in a sampling frame are referred to as the sampling frame **units**. They need not be people. For example, the units may be households, farms, or department stores, and so on.

● EXERCISE 7.4

1 Give two examples in which a sampling frame different from the target population may have to be used.

2 A survey is to be made of listeners' opinions of a local radio station. The sample is to be selected randomly from the electoral registers for the areas covered by the station. Name the target population and the sampling frame. Comment on why the sampling frame used here is unsatisfactory.

7.5 Some Non-Random Sampling Procedures

All the sampling procedures considered so far have involved some element of random sampling. For such procedures statistical theory beyond the level of this book allows one to assess the reliability of any inference that may be made about the population on the basis of the sample responses. For example, such an inference may be about the proportion of the entire population having a particular attribute or opinion, or about the mean value of some variable. This very important advantage is lost when non-random sampling procedures are used. Here we describe some of the most frequently used non-random sampling procedures.

☐☐ PURPOSIVE SAMPLING

An investigator may sometimes use personal judgement in order to choose a sample thought to be representative of the population. Such a procedure is referred to as **purposive sampling** or **judgemental sampling**. For example, in a survey on television viewing the investigator may want a sample made up of so many retired people, so many professional workers, so many manual workers, so many teenagers, and so on. The investigator's aim is to get a sample which is representative of the population of all television viewers. It may well be that the sample is more representative than that obtained by random sampling. The snag is that it is not then possible to make any reliable generalization from the sample responses to the entire population.

One particular form of purposive sampling is known as **quota sampling**. Here an interviewer is told to interview fixed numbers of people of particular types (for example, so many males, so many females, so many teenagers and so on). In most cases the interviewer will make the choice of individuals subject to the quota. The danger is that the interviewer will choose individuals as a matter of convenience rather than randomly. The sample chosen will be in accordance with the quota but is likely to be affected by selection bias.

☐☐ CONVENIENCE SAMPLING

A **convenience sample** is one in which the chosen individuals are those who are easily contacted. The choice of sampled individuals is then a matter of convenience. For example, an interviewer may stand outside a bus or railway station for a sample of people. The results obtained from a convenience sample are likely to be biased and unrepresentative of the population, in which case the results might be very misleading.

The main advantages of convenience sampling are that it is an easy and cheap way of obtaining information from some individuals in a population. For example, suppose that the manufacturer of some product wants customers' opinions on it. A convenient method is for the manufacturer to include a questionnaire with the product for the customer to complete and return to the manufacturer. To encourage a good response, a postage-paid envelope for returning the questionnaire will also be included. This will be much cheaper than paying for professional help in designing and carrying out a sample survey.

EXERCISE 7.5

1 What is meant by (a) quota sampling, (b) convenience sampling? For each type give an example of a situation in which it may be used to advantage.

2 A pupil has been asked to carry out a sample survey on teenagers' opinions on various forms of pop music. The pupil asks all his friends for their opinions. What type of sampling has the pupil used? Do you consider the sample as satisfactory for the purpose of the survey?

7.6 Questionnaires

A set of questions used to obtain information from individuals in a census or sample survey, is called a **questionnaire**. Needless to say, it is important that every question is worded so that the required form of response is obtained. This is not as easy as it might appear. For this reason it is always advisable to try out a questionnaire on some selected individuals before using it in a census or sample survey. This is referred to as a **pretest** of the questionnaire. A pretest usually results in some questions having to be modified.

Here are some useful guidelines to follow when phrasing the questions in a questionnaire.

☐ QUESTIONS SHOULD BE UNAMBIGUOUS

An ambiguous question is of no use since a respondent will not know precisely what is required as a response. Consider a sample survey of houses in which one of the characteristics of interest is the size of a house. To simply ask 'What is the size of your house?' is no good because it is not clear what is meant by size here. Does it mean the total floor area, the volume, the number of floors, the height, or the number of rooms? If what is required is the number of rooms then the question should be phrased as 'How many rooms are there in your house?' But even this is not good enough since it is not clear what is meant by a room. This question appeared in the 1981 census questionnaire, but it was explained that a room did not include a bathroom, a W.C., a room used exclusively for trade or business, or a kitchen if it was less than 2 metres wide. In addition it was stated that a room divided by curtains or portable screens should be regarded as being one room, and a room which was divided by fixed or sliding partitions should be regarded as two rooms. This is a good example to show how careful one has to be when phrasing a question.

In Example 1.2, p. 3 one of the characteristics of the pupils which was of interest was their heights. If the question had been put as 'What is your height?' then the responses might well be in different units of measurement (feet, inches, centimetres, metres) and of different degrees of accuracy. To avoid these possibilities the question was put as 'What is your height in centimetres correct to the nearest centimetre?'

☐ QUESTIONS SHOULD BE APPROPRIATE

It is a waste of time to include questions which will not be answered. This might arise when a question requires technical knowledge not known by the respondent or is of a personal or intimate nature which a respondent will refuse to answer. Considerable care is needed when information of a personal nature is required. (Special techniques have been developed for encouraging more responses to such questions, but these will not be described here.)

An interesting phenomenon that has been observed from data obtained in U.K. censuses is that there appear to be more married females than married males. You can check this by referring to the appropriate table in the *Abstract of Statistics*. Some explanations that have been suggested for this anomaly are that some men were abroad at the time of the census and that some wives who were separated from their husbands rightfully responded that they were married, but their husbands, for whatever reason, responded that they were not married!

☐ AVOID LEADING QUESTIONS

A **leading question** is one which is likely to lead a respondent to give a particular response. For example, asking the question 'You don't approve of taking drugs, do you?' is inviting a 'No' response. Rephrasing it as 'You do approve of taking drugs, don't you?' is inviting a 'Yes' response. A more neutral wording for the question is 'Do you approve or disapprove of taking drugs?' A further point worth making about such a question is that you should allow for the possibility that an individual has not formulated an opinion on the subject. This is done by offering the individual any one of the three responses 'Approve', 'Disapprove', 'No opinion'. A listing of all forms of response that are acceptable is advisable. For example, in a survey of viewers' opinions of some new television programme the alternative responses to offer could be 'Seen and liked', 'Seen but no opinion as yet', 'Seen and not liked', and 'Not seen'.

☐☐ AN EXAMPLE OF A QUESTIONNAIRE

Consider a questionnaire designed to obtain information on the holidays taken last summer by the pupils in a school.

The first part of the planning exercise is to make a list of the sort of information we would like to have. Some possibilities are as follows:

(1) The country in which the pupil holidayed.
(2) The method of travel used.
(3) Alone or with others.
(4) Type of accommodation.
(5) Sex and age of the pupil responding.

What is a holiday? It is a bit ambiguous so we had better say what we mean. Let us agree that a holiday should involve a period of at least 3 nights away from home. That seems clear enough. But what if a pupil has had more than one holiday? To cover this, let us agree that if a pupil had more than one holiday, only the holiday of longest duration should be included in the response.

We can now make a first draft of our questionnaire.

QUESTIONNAIRE ON HOLIDAYS TAKEN LAST SUMMER

Please complete the following questionnaire and return it to (*Name or Location*) by (*Date*).

To ensure confidentiality, you are **not** required to give your name.
(*This last sentence is worth including to encourage more people to respond.*)

1 Your sex.   MALE/FEMALE
2 Your age on your last birthday. . . . .
3 Did you go away on holiday last summer?- YES/NO
   NOTE: A HOLIDAY HERE MEANS ONE IN WHICH YOU SPENT **AT LEAST THREE** NIGHTS AWAY FROM YOUR HOME.
   If NO: ignore the remaining questions and return the questionnaire as requested above.
   If YES: please answer the following questions about your holiday.

*If you had more than one holiday (as defined above) please answer the questions for the holiday of the **longest duration**.*

4 Name the country in which you holidayed. . . . .

5 How many nights were you away from home on the holiday? . . . .

6 Place a tick (√) alongside each mode of travel you used to get to and from your holiday destination:
Bus   Aeroplane   Other
Train Ferry   (please state)
Car   Hovercraft/Hydrofoil

7 Indicate by a tick (√) who accompanied you on your holiday:
No one   Boyfriend(s)
One parent   Girlfriend(s)
Both parents   Other (please state)

8 Indicate by a tick (√) the type of accommodation you had on your holiday:
Hotel   Caravan
Guest house   Camp/Tent
Apartment   Friend's house
Other (please state)

☐☐ OPINION POLLS

A sample survey on people's opinions is often referred to as an **opinion poll**. One problem that arises in an opinion poll as opposed to other forms of sample surveys is that opinions may change over time. Polls on how people would vote if there were an election are held regularly. It is often the case that polls conducted independently at about the same time give very different percentages for the votes for the same political parties. This is not surprising when we realise that different samples of voters have been used. Such differences are said to be due to **sampling error**. What is surprising is that the results obtained in opinion polls should so often be very different from the results of the actual election. Ruling out the possibility that the samples used were biased in some way, the differences may well arise because many voters change their minds between the time of the poll and the actual election (possibly influenced by the results of a poll). The differences may also arise because of the actual votes cast by those who responded 'Don't know' in a poll. Predictions made on radio and television *during* an election are usually more successful because they are based on a sample of votes that have already been cast.

● EXERCISE 7.6

1 What is meant by sampling error?
2 Give an example of a leading question.
3 Comment critically on each of the following questions that are to appear in questionnaires.
(a) Do you have a car? YES/NO (delete as appropriate). If YES, is it an expensive model?
(b) How many times during the past week did you take a bath or a shower?
(c) How many good teachers are there in your school?
(d) What is your weight?
(e) Do you disagree with the view that the town centre should not be closed to traffic on a Saturday?
(f) How many brothers and sisters do you have?
(g) Do you often go to the cinema?
4 What is meant by a pretest of a questionnaire and why is it important?

✳ Projects

Conducting a sample survey is the only way of experiencing the problems that can arise. Here are some examples of surveys for you to do, but feel free to think up alternatives that may be of more interest to you. Remember to pretest your draft questionnaire before using it in the survey. You will need to identify the target population and give careful consideration to your sampling procedure. Having obtained the results of any survey, present the data by means of tables and diagrams, use the techniques described in earlier chapters to calculate appropriate summary measures, and write a brief report on any conclusions you draw.

1 *Carry out the survey discussed on p. 83. First pretest the questionnaire on a small number of pupils and modify the questions if necessary. You may also like to add some more questions to the questionnaire to obtain information on some aspects of a holiday that have not been included.*
2 *Carry out a survey on the pupils in your school to obtain information on weekly pocket money, part-time earnings, how and on what they spend their money, savings accounts, and any other appropriate characteristic.*
3 *Conduct a sample survey on cars driven by ladies in your locality. The cars may be classified by make, type, colour, engine capacity in cc, and*

so on. Also obtain information on the number of miles she has driven during the past year (month), whether she is the only driver of the car, and the type of insurance she has (third party or comprehensive).

4 It is well known that in any year more males are born than females, and that females generally outlive males. Verify these statements by referring to a publication giving the age distributions of males and females at some point in time in a district, county, or country. (One such source for the U.K. population is the annual Abstract of Statistics.)
(a) For any one year extract the age distributions of males and females. For each age or age-group calculate the ratio of the number of males to the number of females. Describe any pattern that is shown by these ratios across the entire age range. Estimate the age when the numbers of males and females are about equal.
(b) Repeat (a) using the age distributions for different years. Comment on the estimated ages (for different years) when the numbers of males and females are about equal.

Review Exercises Chapter 7

1 Name two types of non-random sampling procedures.
2 Explain what is meant by a random sample from a population.
3 Suggest a method for choosing a random sample from each of the following target populations:
(a) The pupils attending a particular school,
(b) The sweets in a box of chocolates in which there are 36 sweets arranged in two layers with each layer being a rectangular array of 6 sweets per row and 3 sweets per column.
(c) The trees of a particular variety in a wood.
(d) The daises growing in a large lawn.
4 Give an example in which stratified random sampling is preferable to random sampling.
5 Give an example in which systematic sampling might be used. State the conditions under which such a sample might not be representative of the population.
6 Give an example of a leading question used in a questionnaire.
7 A local weekly newspaper invites its readers to send in their opinions on a proposal to close the public library on Saturdays. State why the responses obtained may not be representative of the opinions of all the local people.
8 Use a table of random digits (or a calculator or computer) to obtain a random sample of:
(a) 10 individuals from a list of 50 individuals,
(b) 15 individuals from a list of 150 individuals.
9 The following 20 numbers were extracted from a table of random digits:

38, 66, 51, 96, 02, 78, 02, 91, 45, 71,
01, 15, 31, 74, 16, 17, 37, 83, 24, 79,

Using these and the remainder method obtain a random sample of 14 numbers from 01 to 40 inclusive.
10 Describe how you would plan a survey to investigate the shopping habits of the inhabitants of your locality. In particular, the survey is to find out on which days they shop, which shops they choose, and whether advertisements of special offers influenced their choice. Your description should include answers to such questions as: What is the target population? What is the sampling unit? What sampling procedure should be used? What questions should be asked?

8

SOME OTHER SUMMARY MEASURES

"Science is built of facts
the way a house is built of bricks;
but an accumulation of facts
is no more science than a pile of bricks
is a house."

Henri Poincare, La Science et l'hypothese (1902)

8.1 Weighted means

○ EXAMPLE 8.1

A greengrocer sells three varieties (A, B, C) of apples. Apples of variety A are sold at a profit of 12 pence per kilogram, those of variety B at a profit of 15 pence per kilogram, and those of variety C at a profit of 21 pence per kilogram. Calculate the mean profit per kilogram of apples sold.

Here, the variable is profit in pence per kilogram of apples sold and it has three possible values, namely 12, 15 and 21. The mean of these three values is $(12+15+21)/3 = 16$ pence. Does this mean that the greengrocer's average profit is 16 pence per kilogram? The answer is only yes if the three values are of equal importance.

(1) Suppose that on a certain day the greengrocer sold 5 kg of each variety. Then the greengrocer's total profit from the sale of the 15 kg of apples is equal to:

$$(5 \times 12)+(5 \times 15)+(5 \times 21) = 240 \text{ pence},$$

and the mean profit per kilogram sold that day is equal to:

$$\frac{240}{15} = 16 \text{ pence}.$$

(2) Now suppose that on another day the greengrocer sold 8 kg of variety A, 5 kg of variety B, and 2 kg of variety C. On that day the total profit made from the sale of the 15 kg of apples is equal to:

$$(8 \times 12)+(5 \times 15)+(2 \times 21) = 213 \text{ pence},$$

and the mean profit per kg sold that day is then equal to:

$$\frac{231}{15} = 14.2 \text{ pence}.$$

This situation provides an example of a **weighted mean**. The factors by which we multiply the variable values when determining the mean are generally referred to as the **weights** associated with the values. In (1) above the variable values had equal weights, namely 5 (or 5/15 = 1/3), while in (2) the variable values had the weights 8, 5, and 2 respectively (or 8/15, 5/15, and 2/15).

We have already met the principle of a weighted mean in Chapter 4 when calculating the mean of a discrete frequency distribution, the weights associated with the variable values being the frequencies of their occurrence.

○ EXAMPLE 8.2

A manufacturer produces a particular type of article at each of three factories. The production cost per article and the number of articles produced per week at the three factories are given in Table 8.1.

Table 8.1

Factory	Cost per article produced	No. of articles produced per week
I	£2.50	7800
II	£3.06	5600
III	£2.75	6200

Calculate, to the nearest penny, the (weighted) mean production cost per article over all three factories combined.

From the table the total number of articles produced in a week at the three factories is:

$$7800 + 5600 + 6200 = 19\,600,$$

at a total cost of:

$$£((2.50 \times 7800) + (3.06 \times 5600) + (2.75 \times 6200)) = £53\,686.$$

Hence the mean production cost per article is:

$$\frac{£53\,686}{19\,600} = £2.74,$$

to the nearest penny.

In this example the weights associated with the variable values (cost per article produced at the three factories) are the numbers of articles produced per week.

Note that for convenience we could have divided each of the weights by 100 to produce new weights of 78, 56 and 62 respectively. The same answer would have been obtained by using these weights.

In general, if the variable values are $x_1, \ldots x_k$, with associated weights $w_1, \ldots w_k$ respectively, then the weighted mean (WM) is given by the formula:

$$\text{WM} = \frac{w_1 x_1 + \ldots + w_k x_k}{w_1 + \ldots + w_k}$$

In Examples 8.1 and 8.2 the weights to use were fairly obvious. Situations do arise in which determination of the appropriate weights can be quite complicated. For example, the weights used in determining the General Index of Retail Prices (a measure of the monthly changes in the levels of the prices of the commodities and services purchased by households in the U.K.) are based on the pattern of households' expenditures found in family expenditure surveys.

○ EXAMPLE 8.3

Table 8.2 shows the costs and their weights of five related commodities. Calculate the weighted mean cost of (1) commodities A, B and C, (2) all five commodities.

Table 8.2

Commodity	A	B	C	D	E
Cost (£)	5	2	8	6	9
Weight	4	6	2	2	1

(1) The total weighted cost of A, B and C is:

$$£((5 \times 4) + (2 \times 6) + (8 \times 2)) = £48,$$

and the sum of the associated weights is $4 + 6 + 2 = 12$. Hence the weighted mean cost of A, B and C is equal to:

$$\frac{£48}{12} = £4.$$

(2) The total weighted cost of all five commodities is:

$$£((5 \times 4) + (2 \times 6) + (8 \times 2) + (6 \times 2) + (9 \times 1)) = £69,$$

and the sum of the associated weights is $4 + 6 + 2 + 2 + 1 = 15$. Hence the weighted mean cost for all five commodities is:

$$\frac{£69}{15} = £4.60.$$

● EXERCISE 8.1

1 A certain examination consists of three papers of durations 2, $2\frac{1}{2}$ and 3 hours respectively. Weighting each paper by its duration, calculate the weighted mean mark of a candidate whose marks on the papers were 68, 80 and 58 respectively.

2 A Statistics examination consists of a project and one theoretical paper. The project and the paper are each marked out of 100. The final examination mark is obtained by giving the project mark a weighting of 40% and the paper mark a weighting of 60%.
(a) Determine the final mark of a candidate whose project mark was 79 and whose paper mark was 54.
(b) Also determine the project mark of a candidate whose paper mark was 66 and whose final mark was 76.

3 Table 8.3 shows the cost of each of four commodities and the weight to be associated with each cost. Calculate the weighted mean cost.

Table 8.3

Commodity	A	B	C	D
Cost (£)	1·15	1·40	1·20	1·30
Weight	40	50	30	80

4 An office employs twice as many typists as secretaries. A typist is paid £56 per week and a secretary is paid £62 per week. Find the weighted mean weekly wage for persons employed at the office as either a typist or a secretary.

5 A garage sells three grades of petrol. Table 8.4 shows the price per litre (in pence) of each grade and the relative amounts of the grades that are sold at the garage. Calculate the weighted mean price per litre of petrol sold at the garage.

Table 8.4

Grade	2-star	3-star	4-star
Price per litre (p)	40·1	41·3	42·0
% of sales	33	19	48

8.2 Simple Index Numbers

Consider a variable whose value changes over time, for example, the cost of some commodity, the average weekly earnings of manual employees in an industry, consumers' annual expenditure on durable goods, the annual output of some manufacturing industry, vehicle kilometres travelled annually.

For a clearer indication of the pattern of movement of the value of such a variable over time it is customary to choose an appropriate time point (a particular year for a variable observed annually) as a **base** and to express the values of the variable at other points in time (years) as percentages of its value at the base time-point. These percentages are referred to as **relatives** and are the simplest form of what are known generally as **index numbers**. Observe that the value of the relative (or index number) at the base time-point is always 100.

○ **EXAMPLE 8.4**

Table 8.5 shows the price per tonne of a certain raw material for each of the years 1983, 1984 and 1985. Calculate the price relatives for 1984 and 1985 using 1983 as the base year.

Table 8.5

Year	1983	1984	1985
Price per tonne (£)	14·00	15·40	16·10

With 1983 as the base year the price relative for 1984 is the actual price in 1984 expressed as a percentage of the price in 1983. This is given by:

$$\frac{15.40}{14.00} \times 100 = 110.$$

Similarly, the price relative for 1985 with base year 1983 is:

$$\frac{16.10}{14.00} \times 100 = 115.$$

○ **EXAMPLE 8.5**

The index numbers for total annual outputs of a certain manufacturing industry over the period from 1982 to 1985 are given in Table 8.6.

Table 8.6

Year	Index number (output)
1982	100
1983	110
1984	105
1985	120

(1) Given that the output in 1983 was 8.6 million tonnes, determine the output in 1984 to the nearest tenth of a million tonnes.

(2) If the base year is changed to 1983 find the new index numbers, to the nearest integer values, for the years 1982 to 1985.

(3) If the output in 1986 is 15% more than that in 1985 find the index number for the output in 1986 relative to 1982 as the base year.

(1) Here we need to calculate the output in 1984 given that the output in 1983 was 8.6 million tonnes. From the table the ratio of the 1984 output to the 1983 output is 105/110. Hence the output in 1984 is equal to:

$$\frac{105}{110} \times 8.6 = 8.2 \text{ million tonnes,}$$

to the nearest tenth of a million tonnes.

(2) If 1983 is to be taken as the new base year then the index number for 1983 has to be modified from 110 to become 100. This is done by multiplying it by 100/110. This is the multiplying factor to apply to all the old index numbers in order to obtain the new index numbers with 1983 as the base year. The new index numbers are thus, to the nearest integer:

1982: $(100/110) \times 100 = 91$,
1983: 100,
1984: $(100/110) \times 105 = 95$,
1985: $(100/110) \times 120 = 109$.

(3) Since the 1986 output is 15% more than the 1985 output, it is 115% of the 1985 output. The index number for the 1985 output with 1982 as base year is 120, so the corresponding index number for 1986 is given by 115% of 120, which is:

$$\frac{115}{100} \times 120 = 138.$$

● **EXERCISE 8.2**

1 Table 8.7 shows the average cost of a house in a particular region at the beginning of each of several years. Express these average costs as index numbers to the nearest integer, using 1960 as the base year.

Table 8.7

Year	Average cost (£)
1960	2330
1965	3420
1970	4540
1975	10 200
1980	21 540
1985	32 810

2 The costs per kilogram of a certain metal in 1983, 1984, and 1985 were £30, £36 and £40 respectively.
(a) Calculate the metal cost relative for (i) 1984 using 1983 as the base year, (ii) 1983 using 1985 as the base year.
(b) If, with respect to 1983 as base year, the metal cost relative in 1986 was 130, find the cost per kilogram of the metal in 1986.

3 The price relatives of a certain commodity in the years 1982 to 1985, using 1980 as base year, are shown in Table 8.8.

Table 8.8

Year	1982	1983	1984	1985
Price relative	110	105	125	120

(a) If the commodity cost £2 in 1980 find its cost in 1984.
(b) If the price of the commodity in 1986 is 10% higher than it was in 1985 find the price relative for 1986 using 1980 as the base year.

4 A certain index number for 1982 was 60% higher than that for 1981. For 1983 the index number was 50% lower than that for 1982. Find the values of the index numbers for 1981 and 1982 using 1983 as the base year.

5 An index number for 1980, with 1970 as the base year, is 120, while the same index number for 1980, with 1975 as the base year, is 80. Find the index number for:
(a) 1975 with 1970 as the base year,
(b) 1970 with 1975 as the base year.

6 With respect to 1983 as the base year, the index numbers of the total cost of consumer goods for the years 1984 and 1985 were 104 and 108 respectively. Find, to the nearest integer, the new values of the index numbers for 1983 and 1984 if 1985 is taken as the base year.

8.3 Weighted Index Numbers

In the previous section we considered the movement over time of the value of just one variable. Many economic quantities of interest are, in fact, dependent on the values of two or more variables. Given the values of such variables and their associated weights at any given point in time, the weighted mean value is often taken as a measure of the quantity of interest. The weighted means at several time points may then be replaced by index numbers, as in the previous section, to provide a clearer picture of the pattern of change over time of the quantity. Such index numbers are referred to as **weighted** (or **composite) index numbers**. For example, the General Index of Retail Prices (RPI) is based on a weighted mean of the prices of over 600 separate goods and services; the weights are chosen to reflect the spending habits of an 'average' household. As another example, the Financial Times SE Index is based on a weighted mean of the values of the shares of the top 100 companies, but unlike other index numbers it is indexed with respect to the value of 1000 at the base time-point (1 January 1984).

When calculating weighted index numbers it is often the case that the weights are held constant at their values for the base time-point. Since the weights (for example, quantities purchased) may change dramatically over a long period of time it is only realistic to use weighted index numbers with fixed weights over relatively short periods of time.

○ **EXAMPLE 8.6**

Table 8.9 gives information on three raw materials used in a production process for the years 1983, 1984 and 1985. Taking 1983 as the base year calculate index numbers for the overall costs of the three materials in the three years.

Using the weights given in the third column of the table (number of units purchased in 1983), the weighted mean overall costs per unit of the three materials in the three years are found as follows:

1983:
$$\frac{(5800 \times 3.20)+(3500 \times 0.80)+(700 \times 10.00)}{5800+3500+700}$$
$$=\frac{28360}{10000}=£2.836$$

1984:
$$\frac{(5800 \times 3.60)+(3500 \times 1.20)+(700 \times 12.00)}{10000}$$
$$=£3.348,$$

1985:
$$\frac{(5800 \times 3.80)+(3500 \times 1.50)+(700 \times 14.20)}{10000}$$
$$=£3.723.$$

Now taking 1983 as the base year the index numbers for the overall costs are, to one decimal place:

Table 8.9

Material	Unit	No. of units purchased (1983)	Cost per unit (£) 1983	1984	1985
A	kilogram	5800	3·20	3·60	3·80
B	litre	3500	0·80	1·20	1·50
C	barrel	700	10·00	12·00	14·20

1983: 100,
1984: (3.348/2.836) × 100 = 118.1,
1985: (3.723/2.836) × 100 = 131.3.

● **EXERCISE 8.3**

1 Table 8.10 gives the index numbers, with respect to 1980 as the base year, of the industrial outputs of consumer goods in 1985 together with the associated weights. Calculate a weighted index number for the overall output of consumer goods in 1985 with 1980 as the base year, giving your answer to one decimal place.

Table 8.10

Goods	Index number	Weight
Cars	100·5	11
Other durables	105·9	37
Clothing and footwear	103·5	33
Food, drink and tobacco	102·3	89
Other	107·4	75

2 Table 8.11 shows the quantities of four commodities that were purchased by a farmer in 1983, and the costs of these commodities in the years 1983, 1984 and 1985. Calculate weighted index numbers, to one decimal place, for the overall costs of these commodities, using 1983 as the base year.

3 A medicinal mixture contains four ingredients apart from water. The masses, in grams, of these ingredients in a bottle of the mixture and their costs per gram, in pence in 1980 and 1985 are given in Table 8.12. Taking 1980 as the base year,

Table 8.11

Commodity	Unit	No. of units purchased (1983)	Cost per unit (£) 1983	1984	1985
A	kilogram	120	0·45	0·46	0·48
B	tonne	50	8·60	9·00	9·30
C	kilogram	60	0·80	0·70	0·66
D	metre	100	1·20	1·30	1·40

Table 8.12

Ingredient	Mass (g)	Cost per gram (p) 1980	1985
1	12	4·2	4·8
2	5	1·2	2·0
3	6	1·6	1·8
4	10	2·4	2·8

calculate to one decimal place an index number for 1985 for the overall cost of the ingredients per bottle of the mixture.

4 *Table 8.13 gives the index numbers, with respect to 1974 as the base year, of the prices of different varieties of pig meat in 1982, 1983 and 1984, together with the associated index weights. For each year calculate the weighted means of the index numbers. Express the weighted means as index numbers, to the nearest integer, using 1982 as the base year.*

Table 8.13

Variety	1982	1983	1984	Weight
Pork	226·6	266·8	232·8	6
Bacon	231·9	237·2	238·9	6
Ham (cooked)	215·0	225·2	233·1	2

8.4 The Geometric Mean*

Hitherto we have made extensive use of the mean (or, more strictly, the arithmetic mean) as a measure of the average value of a set of numbers. However, the mean is not an appropriate measure for averaging ratios, percentages, or rates of change over time, as illustrated by the following example.

○ **EXAMPLE 8.7**

Suppose a person invested £10 000 in the shares of a certain company on 1 January 1980, and that the values of those shares on 1 January of the succeeding four years were £11 000, £12 650, £13 915, and £16 698 respectively. What was the average annual percentage increase in the value of the shares over the four year period?

To find the average percentage increase per year we first need to calculate the yearly percentage increases. Since the initial £10 000 had increased in value to £11 000 a year later, the increase in value is £1000 and the percentage increase for that year is:

$$\frac{1000}{10\,000} \times 100 = 10\%.$$

Similarly, the percentage increase in the value from 1981 to 1982 is:

$$\frac{12\,650 - 11\,000}{11\,000} \times 100 = 15\%,$$

that from 1982 to 1983 is:

$$\frac{13\,915 - 12\,650}{12\,650} \times 100 = 10\%,$$

and that from 1983 to 1984 is

$$\frac{16\,698 - 13\,915}{13\,915} \times 100 = 20\%.$$

The average of these four percentage increases is:

$$\frac{10+15+10+20}{4} = 13.75.$$

Does this mean that if the value of the shares increased at a constant 13.75% per year then the value of the shares in four years time would have been equal to their actual value of £16 698? Let's find out by applying a constant yearly increase of 13.75% for four successive years. In January of each succeeding year the value of the shares in this case would be as follows:

1981: 113.75% of £10 000 = £11 375,
1982: 113.75% of £11 375 = £12 939.063,
1983: 113.75% of £12 939.063 = £14 718.184,
1984: 113.75% of £14 718.184 = £16 741.934,

which is somewhat higher than the actual value of £16 698. Thus the average of the yearly percentage increases does not lead to the same final value that was realised from the initial investment.

To determine the constant annual percentage increase that will yield the same final value of the shares in 1984 we need to use the **geometric mean**.

[91

The geometric mean (GM) of a set of numbers $x_1, \ldots x_n$ is defined to be the nth root of their product; that is:

$$GM = \sqrt[n]{(x_1 \times x_2 \times \ldots \times x_n)}.$$

Returning to our present example, we first express the value of the shares in any year as a percentage of their value in the preceding year. These percentages are as follows.

1981 value as % of 1980 value:
$(11\,000/10\,000) \times 100 = 110\%$,
1982 value as % of 1981 value:
$(12\,650/11\,000) \times 100 = 115\%$,
1983 value as % of 1982 value:
$(13\,915/12\,650) \times 100 = 110\%$,
1984 value as % of 1983 value:
$(16\,698/13\,915) \times 100 = 120\%$.

Now take the geometric mean of these four percentages, which is given by:

$$GM = \sqrt[4]{(110 \times 115 \times 110 \times 120)} = 113.6753,$$
(to 4 d.p.),

corresponding to an average yearly increase of 13.6753%. Applying this constant yearly percentage increase over four years, the final value of the shares would be:

£10 000 × 1.136753 × 1.136753 ×
1.136753 × 1.136753 = £16 697.999,

which agrees with the actual value on allowing for the fact that the geometric mean was calculated to four decimal places only. It follows that finding the average yearly percentage increase using the geometric mean is more appropriate than taking the arithmetic mean of the individual yearly percentage increases; the calculated value of the shares will then be close to their actual value.

○ **EXAMPLE 8.8**

Calculate the geometric mean of each of the following sets of numbers, giving each answer correct to the nearest whole number.

(1) 125, 136, 118, 129, 146,
(2) 34, 46, 21, 32, 58, 138.

(1) For the five numbers 125, 136, 118, 129, 146 the geometric mean is given by:

$$GM = \sqrt[5]{(125 \times 136 \times 118 \times 129 \times 146)}$$
$$= \sqrt[5]{(3.7781004 \times 10^{10})}.$$

The product has been shown in the form in which it was displayed by the calculator used. With this product still in the display of the calculator the fifth root may be obtained by pressing the following sequence of keys: $\boxed{\sqrt[x]{y}}\,\boxed{5}\,\boxed{=}$.

The resulting number, 130.45319, displayed by the calculator is the required fifth root. Thus to the nearest whole number the geometric mean is 130.

(2) For the six numbers 34, 46, 21, 32, 58, 138 the geometric mean is given by:

$$GM = \sqrt[6]{(34 \times 46 \times 21 \times 32 \times 58 \times 138)}.$$

Following the calculator procedure described above, but replacing $\boxed{5}$ by $\boxed{6}$ (since we now want the sixth root), we find that the geometric mean is given by:

$$GM = 45.097 \quad \text{(to 3 d.p.)},$$

which is 45 to the nearest whole number.

It may be shown that for any set of positive numbers the geometric mean is always less than the arithmetic mean. Also, the geometric mean is less affected than the arithmetic mean by the presence of an abnormally large value. To illustrate the former it is left as an exercise to show that the arithmetic means of the numbers in (1) and (2) of Example 8.8 are 130.8 (as compared with GM = 130.45), and 54.833 (as compared with GM = 45.097) respectively. Note also that in (2) the presence of the relatively large value 138 has less effect on the value of the geometric mean as compared with the arithmetic mean.

The geometric mean is always appropriate for averaging ratios and rates of change. Since index numbers (simple or weighted) are ratios measuring rates of change, an average index number is often calculated by taking the geometric mean instead of the arithmetic mean, which we did in Section 8.3 when calculating a weighted index number. Given k numbers $x_1, x_2, \ldots x_k$, having associated weights $w_1, w_2, \ldots w_k$, with

$$w_1 + w_2 + \ldots + w_k = n,$$

the **weighted geometric mean** (WGM) is given by:

$$WGM = \sqrt[n]{(x_1^{w_1} \times x_2^{w_2} \times \ldots \times x_k^{w_k})}$$

○ **EXAMPLE 8.9**

Table 8.14 shows the retail prices in pence of three items of food in 1980 and 1985, together with the

Table 8.14

Item	Price (p) 1980	Price (p) 1985	Weight
A	10	12	1
B	12	18	3
C	20	25	2

associated weights. Calculate the price relative for each item in 1985, with 1980 as the base year. Hence calculate a weighted index number for the 1985 overall price, with respect to 1980 as the base year, by determining the weighted geometric mean of the price relatives. Give the answer to the nearest whole number.

With 1980 as the base year the price relatives of the items in 1985 are:

A: $(12/10) \times 100 = 120$,
B: $(18/12) \times 100 = 150$,
C: $(25/20) \times 100 = 125$.

Since the sum of the weights is $1+3+2=6$, the weighted geometric mean (WGM) of these price relatives is given by:

$$\text{WGM} = \sqrt[6]{(120^1 \times 150^3 \times 125^2)} = \sqrt[6]{(6.328125 \times 10^{12})}$$
$$= 136.002 \quad \text{(to 3 d.p.)},$$
$$= 136 \text{ to the nearest whole number.}$$

● **EXERCISE 8.4**

1 Calculate the geometric mean of each of the following sets of numbers and verify that its value is less than the arithmetic mean of the numbers.

(a) 110, 124, 136,
(b) 2.56, 4.58, 9.26, 11.34.

2 The annual profits made by a company over the period 1980 to 1983 are shown in Table 8.15, together with these profits expressed as index numbers using 1981 as the base year.
(a) Determine the values of a and b.

Table 8.15

Year	1980	1981	1982	1983
Profit (£m)	2.31	2.75	2.42	a
Index number	b	100	88	112

(b) Calculate the geometric mean of the index numbers for the years 1981, 1982 and 1983, giving your answer to the nearest whole number.

3 In January 1980 a man's salary was increased by 8% of his 1979 salary; in January 1981 he received an increase of 10% of his 1980 salary; and in January 1982 he received an increase of 12% of his 1981 salary.
(a) Express the man's salary in each of the years 1980, 1981 and 1982 as a percentage of his salary in the immediately preceding year.
(b) Calculate the geometric mean of these percentages.
(c) Deduce what constant annual percentage increase in salary over the three-year period would result in the man having the same 1982 salary.

4 When a certain plant is grown under controlled conditions its height increases by 20% per month. At the end of April the height of the plant was 15 cm.
(a) Find its height at the end of each month from May to July, inclusive.
(b) Calculate the geometric mean of the heights of the plant at the end of the months April to July.
(c) State why the geometric mean is an appropriate measure of the average height of the plant.

5 Table 8.16 shows the average prices in pounds per kilogram of three types of beverage in the years 1975 and 1985, together with the weights to be associated with these prices.
(a) Determine the price relatives for 1985 with 1975 as the base year.
(b) Calculate the weighted geometric mean of these price relatives, giving your answer to the nearest whole number.

Table 8.16

Year	Price per kg (£)		
	Tea	Coffee	Cocoa
1975	0.32	0.80	0.50
1985	0.96	3.40	2.60
Weight	5	3	2

8.5 Crude and Standardized Rates

☐☐ DEATH RATES

The death rate for people residing in a particular region (which may be a town, a city, a county or a country) in a given year is usually quoted as the number of deaths per thousand people. For example, if in a given year the population of a city was 32 400 and the number of deaths during that year was 405, then the city's death rate for that year was:

$$\frac{405}{32\,400} \times 1000 = 12.5 \text{ per thousand.}$$

Since the population size of a region will vary throughout a year (through deaths, births and migration) it is customary to take the population size to be that near the middle of the year of interest (for example, towards the end of June).

In addition to the death rate for the entire population of a region, we are often also interested in the death rate for people of a particular age or within a particular age-group; this is referred to as an **age-specific death rate**. For example, if the number of people in the age-group 40–59 in a region in a particular year is 6548, and 59 of them die during the year, then the age-specific death rate for the age-group 40–59 in that region is:

$$\frac{59}{6548} \times 1000 = 9.01 \text{ per thousand} \quad \text{(to 2 d.p.)}.$$

The death rate for any region clearly depends on the age distribution of the residents. A region which is largely populated by old people will have a relatively higher death rate as compared with a region having a low proportion of old people. For this reason a death rate calculated as described above is referred to as a **crude death rate**. A more appropriate measure for comparing the death rates of different regions by taking account of their age distributions will be discussed after the following example.

○ EXAMPLE 8.10

Table 8.17 shows the age distributions of two particular towns (A and B) in 1984 together with the numbers of deaths in that year within each age-group. Calculate the age-specific death rates in Town A, and the overall crude death rates for the two towns.

Table 8.17

	Town A		Town B	
Age-group	Population ('000)	No. of deaths	Population ('000)	No. of deaths
0–14	3	30	8	55
15–44	4	26	25	65
45–64	3·5	28	13	150
65+	1·5	63	14	570
Total	12·0	147	60	840

For Town A, the age-specific death rates per thousand people are as follow:

$$\begin{aligned} 0\text{–}14: & \quad 30/3 = 10 \\ 15\text{–}44: & \quad 26/4 = 6.5 \\ 45\text{–}64: & \quad 28/3.5 = 8 \\ 65+: & \quad 63/1.5 = 42. \end{aligned}$$

Using the total entries in the last row of the table, the overall crude death rates per thousand people for the two towns are:

Town A: $147/12 = 12.25$,
Town B: $840/60 = 14$.

☐☐ STANDARDIZED DEATH RATES*

It would appear from the answers obtained in the previous example that for longevity of life Town A may be a better place to live in than Town B, since the death rate for Town A is less than that for Town B. However, as remarked earlier, comparing crude death rates may be unfair if the age distributions in the two towns are very different, and particularly so when the proportions of older people in the two towns are very different. Referring to Table 8.17 we see that the proportions of people in the age-group 65+ in the two towns are:

Town A: $1.5/12 = 0.125$ or 12·5%,
Town B: $14/60 \approx 0.233$ or 23.3%.

Since the proportion of people aged 65+ in Town B is almost double that in Town A it is not surprising that Town B has the higher crude death rate.

For a fairer comparison we need to take account of the difference in the age distributions of the populations in the two towns. This may be achieved by estimating the overall death rate for Town A, say, if its population had the same age distribution as that of Town B. We refer to this as a **standardized death rate** for Town A, the age distribution of Town B being used as the standard. To determine this standardized death rate for Town A we need the age-specific death rates for Town A and the age distribution of Town B. The calculations are illustrated in the following example.

○ EXAMPLE 8.11

Using the information given in Table 8.17 calculate the standardized death rate for Town A, using the age distribution of Town B as the standard.

The age-specific death rates for Town A were calculated in Example 8.10 and are shown in the second column of Table 8.18. The third column of the table shows the age distribution in thousands of people for Town B, which is to be taken as the standard. The fourth column of the Table shows the values of the products of the entries in the second and third columns. Each entry in this fourth column is the estimated number of deaths per thousand people in the corresponding age-group if the number of people in Town A in that age-group had been equal to that in Town B.

Table 8.18

Age-group	Age-specific death rates (A)	Standard population (B)	Estimated no. of deaths
0–14	10	8	80
15–44	6.5	25	162.5
45–64	8	13	104
65+	42	14	588
Total	—	60	934.5

The standardized death rate for Town A with respect to the age distribution in Town B is the ratio of the estimated number of deaths (934.5) to the size in thousands of the standard population (60). So the standardized death rate for Town A is:

$$\frac{934.5}{60} = 15.575 \text{ per thousand.}$$

Recalling that the (crude) death rate in Town B was 14 per thousand, it follows that Town B is, in fact, the healthier town to live in.

It is worth noting that the standardized death rate for Town A, with the age distribution of Town B as standard, is simply the weighted mean of the age-specific death rates for Town A, with weights equal to the age-group population sizes in Town B.

In practice, standardized death rates for regions are usually calculated taking the age distribution of the population of the entire country as the standard. For the age-groups used in Examples 8.10 and 8.11, the age distribution of the population of the U.K. in 1984 was as shown in Table 8.19 (based on Table 1.1 of the 1986 edition of *Social Trends*).

Table 8.19 *Age Distribution of the U.K. Population in 1984*

Age-group	0–14	15–44	45–64	65+
Population (m)	11	24.5	12.6	8.4

○ EXAMPLE 8.12

Using the age distribution in the U.K. in 1984, (given in Table 8.19) as the standard, calculate the standardized death rates for Towns A and B of Example 8.10.

We first need to calculate the age-specific death rates for the two towns. We have already done so for Town A in Example 8.10. Referring to the Town B entries in Table 8.17, the age-specific death rates per thousand people are as follows:

0–14: 55/8 = 6.875,
15–44: 65/25 = 2.600,
45–64: 150/13 = 11.54,
65+: 570/14 = 40.71.

Table 8.20

Age-group	Age-specific death rates Town A	Age-specific death rates Town B	Standard population (m)
0–14	10	6.88	11
15–44	6.5	2.6	24.5
45–64	8	11.54	12.6
65+	42	40.71	8.4
		Total	56.5

Table 8.20 shows the age-specific death rates for Towns A and B, and the age distribution in the standard population, namely that in the U.K. in 1984.

Using the standard population figures as weights, the weighted means of the age-specific death rates of Towns A and B, which are the standardized death rates per thousand people are as follows:

Town A:
$$\frac{(10 \times 11) + (6.5 \times 24.5) + (8 \times 12.6) + (42 \times 8.4)}{56.5}$$
$$= \frac{722.85}{56.5} = 12.79 \quad \text{(to 2 d.p.)},$$

Town B:
$$\frac{(6.875 \times 11) + (2.6 \times 24.5) + (11.54 \times 12.6) + (40.71 \times 8.4)}{56.5}$$
$$= \frac{626.693}{56.5} = 11.09 \quad \text{(to 2 d.p.)}.$$

Since the standardized death rate for Town B is less than that for Town A, the conclusion is, as in Example 8.11, that Town B is the healthier one to live in.

○ EXERCISE 8.5

1 Table 8.21 shows the age distribution of the population of a town in 1982 and the number of

Table 8.21

Age-group	Town population ('000)	No. of deaths	Population of country (millions)
0–4	3	30	3.6
5–29	6	6	20.7
30–44	9	18	11.2
45–59	12	72	9.3
60–74	3.6	90	8.1
75+	2.4	288	3.6

deaths during the year in each age-group. The final column shows the corresponding age distribution for the entire country in millions.
(a) Calculate the crude death rates for (i) the town population, (ii) the people in the town who are aged 30 and over. (b) Calculate the corresponding standardized death rates using the country population as standard.

2 Table 8.22 gives data relating to the population of a particular town in 1981.
(a) Calculate the crude death rate for this town in 1981.
(b) Given that in the country as a whole the percentages of the population falling in the tabulated age-groups in 1981 were 12%, 18%, 56%, and 14% respectively, calculate the standardized death rate.

Table 8.22

Age-group	Population	No. of deaths
0–5	10 000	130
6–15	12 000	10
16–55	30 000	160
56+	8 000	320

3 Table 8.23 shows the populations and the numbers of deaths of infants under 5 years of age in a certain country in each of the years 1975 and 1985.
(a) Calculate the crude death rate for the infant population in each of the two years.
(b) Calculate the standardized infant death rate for 1985 using 1975 as the standard.

Table 8.23

Year	2 years old and under Population	2 years old and under Deaths	Over 2 and under 5 years old Population	Over 2 and under 5 years old Deaths
1975	22 730	980	34 210	169
1985	27 440	491	48 240	47

4 Table 8.24 gives the age distribution and the annual death rate per thousand for each age-group of the adult male population of a town in 1984, together with the corresponding age distribution for all adult males in the country.
(a) Calculate the crude and the standardized death rates (b) The crude and the standardized death rates for adult males in a city in the same year are 2.46 and 4.32 per thousand respectively. State, giving your reason, whether the town or the city has the healthier environment for adult males.

Table 8.24

Age-group	% of town's male population	Death rate (per '000)	% of country's male population
16–34	48	1	38
35–59	38	4	39
60+	14	6	23

☐☐ OTHER RATES

Other rates that may be associated with the population of a region include the birth rate, the fertility rate, the marriage rate, and the unemployment rate. A birth rate is expressed as the number of live births per thousand of the population, or per thousand females of child-bearing age. As was the case for the death rate, the birth rate of a region will be dependent upon the age structure of the population, and it needs to be standardized for comparison purposes. An unemployment rate is usually expressed as the number unemployed per hundred adults.

In addition to rates related to a population there are also other types of rates in common use, as illustrated in Example 8.14 and some of the following exercises.

○ EXAMPLE 8.13

Table 8.25 gives an age distribution of females of child-bearing age together with the number of children given birth by these females in each age-group in a certain region in 1984.

Table 8.25

Age (years)	No. of females	No. of births
15–19	104	4
20–24	90	10
25–29	96	12
30–34	98	7
35–39	86	2
40–49	126	1

(1) Calculate the crude birth rate.
(2) If in the country as a whole the percentages of all females in the tabulated age-groups are 9.1%, 8.1%, 8%, 7%, 4.3%, and 7% respectively, calculate the standardized birth rate using the age-distribution of the entire country as the standard.

(1) The total number of births in the region in 1984 was:

$$4+10+12+7+2+1=36,$$

and the total number of females of child-bearing age was:

$$104+90+96+98+86+126=600.$$

Hence the crude birth rate per thousand females of child-bearing age is:

$$\frac{36}{600} \times 1000 = 60.$$

Research carried out by the National Perinatal Epidemiology Unit in Oxford claims that hospital births may not be safest. These findings run contrary to the policy of successive governments and much of the medical establishment's thinking on the subject. The dangers of unforeseen complications have prompted a policy of encouraging all women to have their babies in hospital, preferably in large units with the full range of obstetric and paediatric services. Indeed, there has been a policy of closure for many smaller units.

The number of home births in 1975 was 19 000. This fell to fewer than 6000 in 1985. Of the 660 000 babies born that year, nearly all were delivered in hospital, and the vast majority (nearly 630 000) in consultant-staffed units at district hospitals.

On the face of it, the decline in baby deaths from 19·2 per thousand in 1975 to 9·8 per thousand in 1985 would seem to support government policy. The mortality rate of 24 per thousand among home deliveries is more than twice that among babies delivered in hospital.

So what data do the researchers have to support their argument? A closer look at the deaths for home-delivered babies is necessary. Most of the apparently high death rate, say the researchers, resulted from women who, already booked to go into hospital, had delivered very fast or very prematurely at home. Here the death rate was 67 per thousand. Mothers who had made no arrangements at all for either a home or hospital delivery also contribute to the high death rate. Almost 1 in 5 of babies born in such circumstances died.

In fact, for those whose delivery at home was planned, the death rate was only 4·1 per thousand. The researchers claim that in cases where there are no foreseen complications, a mother-to-be who chooses to have her baby at home should be able to do so. It is possible, although there are no data to show it, that she and her baby may be at risk since the advantages of greater institutional care also carry with them increased risks of serious cross-infection.

(2) For the standardized birth rate we first need to calculate the age-specific birth rates for the various age groups, which are as follows:

15–19: $(4/104) \times 1000 \approx 38.46$,
20–24: $(10/90) \times 1000 \approx 111.11$,
25–29: $(12/96) \times 1000 = 125$,
30–34: $(7/98) \times 1000 \approx 71.43$,
35–39: $(2/86) \times 1000 \approx 23.26$,
40–49: $(1/126) \times 1000 \approx 7.94$.

For convenience these age-specific birth rates are displayed in Table 8.26 together with the percentages of all females in the country in the listed age-groups.

Table 8.26

Age-group:	Birth rate	% of all females
15–19	38·46	9·1
20–24	111·11	8·1
25–29	125	8
30–34	71·43	7
35–39	23·26	4·3
40–49	7·94	7

The standardized birth rate is then given by finding the weighted mean of the age-specific birth rates with weights as indicated in the last column of the table. Thus the standardized birth rate is:

$$\frac{(38.46 \times 9.1)+(111.11 \times 8.1)+(125 \times 8)+(71.43 \times 7)+(23.26 \times 4.3)+(7.94 \times 7)}{9.1+8.1+8+7+4.3+7}$$
$$= 2905.585/43.5 = 66.80 \quad \text{(to 2 d.p.)},$$

which is 67 per thousand females of child-bearing age, to the nearest integer.

○ EXAMPLE 8.14

A nursery grows plants of two varieties (A and B) both of which are susceptible to a certain disease. Table 8.27 shows the age distribution of the plants of the two varieties, and for each age-group the number of plants that are affected by the disease.

Table 8.27

Age of plants (days)	Variety A No. of plants	Variety A No. diseased	Variety B No. of plants	Variety B No. diseased
0–30	400	22	800	20
31–60	200	4	600	28
61–90	200	6	300	16
91+	200	4	300	9

(1) Find the crude rates of incidence of the disease for the two varieties of plants.
(2) It is known that the susceptibility of a plant to the disease depends upon the age of the plant. Taking this into account, determine which of the two varieties is the more susceptible to the disease.

(1) For Variety A the total number of plants is $400+200+200+200=1000$, the number of diseased plants is $22+4+6+4=36$. Hence the crude rate of incidence of the disease in plants of Variety A is:

$$\frac{36}{1000} \times 1000 = 36 \text{ per thousand plants.}$$

Similarly, for Variety B the total number of plants is $800+600+300+300=2000$. The number of diseased plants is $20+28+16+9=73$. Hence the crude rate of incidence of the disease in plants of Variety B is:

$$\frac{73}{2000} \times 1000 = 36.5 \text{ per thousand plants.}$$

(2) Since the age distributions of the two varieties of plants are different, and susceptibility to the disease is known to depend on age, the crude incidence rates are not appropriate for comparing the susceptibilities of the two varieties. To take account of the different age structures we need standardized rates. Let us take the age distribution of plants of Variety A as the standard. Then the standardized rate of incidence of the disease for plants of Variety A will be equal to the crude rate of incidence, namely 36 per thousand.
To determine the corresponding standardized rate for plants of Variety B we first need to calculate the age-specific rates per thousand plants, which are as follows:

0–30: $(20/800) \times 1000 = 25$,
31–60: $(28/600) \times 1000 \approx 46.67$,
61–90: $(16/300) \times 1000 \approx 53.33$,
91+: $(9/300) \times 1000 = 30$.

Using the age distribution of plants of Variety A as the standard, the standardized incidence rate for plants of Variety B is:

$$\frac{(25\times400)+(46.67\times200)+(53.33\times200)+(30\times200)}{400+200+200+200}$$

$= 36\,000/1000 = 36$ per thousand plants.

Since this is equal to the standardized rate of incidence for plants of Variety A, we conclude that the two varieties are equally susceptible to the disease.

5 Table 8.28 shows the percentages of men of various age-groups that were unemployed in a certain region in a given year. Of all the men in the country aged 18 and over, 25% are in the age-group 18–30, 36% are in the age-group 31–45, 28% are in the age-group 46–59, and the remaining 11% are at least 60 years old. Using the age distribution of men in the country as the standard, calculate the standardized unemployment rate for the region.

Table 8.28

Age-group	18–30	31–45	46–59	60+
% unemployed	3·2	2·6	6·8	22·5

6 Table 8.29 gives data on the number of children given birth by married females in Towns A and B in 1985.

Table 8.29

	Town A		Town B	
Age of married females (years)	No. of births	No. of females	No. of births	No. of females
Under 20	40	1000	40	1000
20–29	320	2000	640	4000
30–39	200	2000	300	3000
40+	40	5000	16	2000

(a) Calculate the crude legitimate birth rate for each town. State why it is apparent from the table that the rate for Town B is substantially greater than that for Town A.
(b) For the entire country the percentages of married females in the tabulated age-groups are 10%, 20%, 20%, and 50% respectively. Use this age distribution as the standard to calculate standardized legitimate birth rates for the two towns. State why the crude and standardized rates for Town A are equal.

7 A gardener sowed carrot seeds in two different types of compost. The seeds had been stored for various lengths of time prior to planting. Table 8.30 shows for each storage time period the number of seeds that were sown and the number that germinated for each type of compost.
(a) Calculate the crude germination rate for each compost.
(b) The germination rate is known to be dependent upon the storage period. Taking this into account, determine which compost has the higher germination rate.

Table 8.30

Storage period (months)	Compost A		Compost B	
	No. sown	No. germinated	No. sown	No. germinated
0–6	30	28	40	28
7–12	80	68	30	22
13–18	90	50	20	15
19+	50	4	10	5

8 Table 8.31 shows the age-distributions of unmarried females (16 and over) in two districts in a certain year, and the numbers of these that got married during that year. For each district calculate:
(a) the crude overall marriage rate,
(b) the standardized overall marriage rate, using as standard the age distribution for the entire country, which is such that of all unmarried females age 16 and over, 60% are in the age-group 16–24 and 30% are in the age-group 25–45.

Table 8.31

	District A		District B	
Age-group	No. unmarried ('000)	No. of marriages	No. unmarried ('000)	No. of marriages
16–24	22	320	25	280
25–45	15	220	10	110
46+	3	40	5	55

Review Exercises Chapter 8

☐ LEVEL 1

1 A Mathematics examination consists of two papers, the first being of duration $1\frac{1}{2}$ hours and the second of duration $2\frac{1}{2}$ hours. Weighting the mark in each paper by its duration, calculate:
(a) the weighted mean mark of a candidate whose marks in the two papers were 88% and 64% respectively,
(b) the mark of a candidate in the first paper given that this candidate had a weighted mean mark of 71% and a mark of 62% in the second paper.

2 In each of the years 1983, 1984, and 1985 a firm purchased four items in the quantities and at the prices shown in Table 8.32.
(a) Calculate the overall cost of the four items in each of the years.
(b) Using 1983 as the base year express these overall costs as index numbers.

Table 8.32

			Price per unit (£)		
Item	Unit	Quantity	1983	1984	1985
1	kilogram	170	4	5	6
2	litre	200	7	7	8
3	hundred	130	12	13	15
4	metre	140	8	7	6

3 In 1983 a factory had 140 employees and its total wages bill was £250 000. In 1984 the factory had 160 employees and its total wages bill was £260 000. Using 1983 as the base year find the 1984 index number for:
(a) the number of employees,
(b) the total wages paid,
(c) the average wage paid per employee.

4 The information given in Table 8.33 relates to four raw materials used in a manufacturing process, the price relatives being with respect to 1980 as the base year.
(a) Calculate to one decimal place in each case (i) the price per tonne of Material B in 1982, (ii) the price relative of Material C in 1984 with 1982 as the base year. (b) Using the weights given in the final column of the table, calculate, to the nearest whole number, a weighted index number for the overall expenditure on the four raw materials in (i) 1982, (ii) 1984 using 1980 as the base year.

Table 8.33

Material	Price per tonne (£) in 1980	Price relatives 1982	1984	Weight
A	14·00	110	115	8
B	3·40	136	130	5
C	69·50	117	94	3
D	58·00	98	100	4

5 With 1970 as the base year the price relative of a particular product in 1975 was 120. With 1975 as the base year the price relative of the same product in 1980 was 150. In 1975 the product cost £4, and in 1985 it cost £10. Calculate:
(a) the index for 1985 using 1975 as the base year,
(b) the index for 1985 using 1970 as the base year,
(c) the cost of the product in 1970.

6 Table 8.34 gives the index numbers of the prices of three categories of beverage in January of each of the years 1982, 1983, and 1984, with respect to the prices in January 1974 as base. The weights to be associated with each category of beverage are also given.
(a) Determine which of the categories of beverage had the largest percentage increase in price from 1982 to 1984.
(b) Calculate, to the nearest whole number, the weighted index numbers for the overall prices of the beverages in 1982, 1983, and 1984, using (i) 1974 as the base year, (ii) 1982 as the base year.

Table 8.34

	Price index			
Beverage	1982	1983	1984	Weight
Tea	330·0	330·7	385·3	3
Coffee and cocoa	318·2	346·6	387·9	3
Soft drinks	314·0	316·2	331·9	5

7 With 1975 as the base year the index number for the number of tonnes of a certain commodity that was sold in 1985 was 120, while the index number for the profit per tonne sold was 115. Find the index number for the total profit from the sale of this commodity in 1985 using 1975 as the base year.

8 Table 8.35, extracted from the 1985 Edition of *Transport Statistics*, gives the index numbers,

with 1980 as the base year, for expenditures on various forms of public transport in the U.K. during the years 1980–1984.
(a) For which form of transport was the percentage increase in expenditure from 1980 to 1984 (i) the least, (ii) the greatest?
(b) Express the index number for rail expenditure in each of the years from 1981 to 1984 as a percentage of the expenditure in the immediately preceding year.
(c) State the year when the percentage increase was greatest.

Table 8.35

Form of transport	Index number				
	1980	1981	1982	1983	1984
Rail	100	109	117	124	127
Bus & coach	100	114	129	134	139
Air	100	105	114	128	121
Sea	100	111	119	129	143

9 The retail price indices and the associated weights of four commodities in a certain year are shown in Table 8.36. Given that the weighted (arithmetic) mean index is 103, find the value of x.

Table 8.36

Commodity	A	B	C	D
Price index	104	101	102	107
Weight	2	x	4	3

☐ **LEVEL 2**

10 Table 8.37 shows the average prices in pence per kilogram of three products in the years 1975 and 1985.
(a) Calculate the price relatives for 1985 using 1975 as the base year. (b) Calculate the geometric mean of these price relatives giving your answer correct to one decimal place.

Table 8.37

Product	Average price	
	1975	1985
A	26	31
B	42	49
C	56	60

11 £1000 was deposited in a bank account on 1 January 1980. Interest at 5% per annum is added to the amount on deposit each succeeding 1 January.
(a) Draw up a table to show the amounts on deposit on 1 January of each of the years from 1980 to 1983.
(b) Calculate the arithmetic mean and the geometric mean of the amounts on deposit in the years 1980, 1981, 1982 and 1983.
(c) State, with your reason, which of these means is the more appropriate measure of the average yearly amount on deposit.

12 Table 8.38 shows the total price and the tax (duty and VAT) in pence per gallon of petrol in January of each of the years from 1982 to 1985.

Table 8.38

Year	1982	1983	1984	1985
Total price (p)	159	167	183	189
Tax (p)	83·57	92·43	97·97	102·65

(a) Using 1982 as the base year, express as index numbers to one decimal place (i) the total prices for the tabulated years, (ii) the taxes for the tabulated years, (iii) the total prices excluding tax for the tabulated years.
(b) Express the increase in the total price from 1982 to 1983 as a percentage of the total price in 1982, the increase in the price from 1983 to 1984 as a percentage of the total price in 1983, and the increase in the total price from 1984 to 1985 as a percentage of the total price in 1984.
(c) Calculate the geometric mean of the three percentage increases in (b), giving your answer to one decimal place.

13 The retail price indexes and the associated weights for three commodities in a certain year are shown in Table 8.39. Obtain a composite retail index for the three commodities by calculating the weighted geometric mean of the individual price indexes, giving your answer to the nearest whole number.

Table 8.39

Commodity	I	II	III
Price index	110	122	98
Weight	2	2	1

14 For a certain town the crude death rate in a particular year was 14.7 per thousand. Of the people living in the town that year, 75% were

[101

under 65 years old. Given that the age-specific death rate for the people in the town who were aged 65 and over was four times that for those aged under 65, calculate the age-specific death rates for those people (a) under 65, (b) 65 and over.

15 Table 8.40 shows an age distribution of the population of a certain town in 1984, together with the age-specific death rates and the numbers of deaths in the various age groups.
(a) Calculate the crude death rate for the town.
(b) Using the age distribution of the U.K. population in 1984 (see Table 8.19, p. 95) as the standard, calculate the standardized death rate for the town in 1984.

Table 8.40

Age (years)	0–14	15–44	45–64	65+
Death rate (per '000)	1·8	1·6	14·0	122·3
No. of deaths	21	56	338	1284

16 The death rates from a certain disease for male employees in Industry A are 0.3 per thousand per annum for those aged under 50 and 2.3 per thousand per annum for those aged 50 and over.
(a) Given that 45% of the males employed in Industry A are aged 50 and over, calculate the crude overall death rate per annum for males in Industry A.
(b) For male employees in Industry B the crude overall death rate from the same disease is 0.9 per thousand per annum. For these employees the death rate for those aged under 50 is 1.8 per thousand per annum, while for those aged 50 and over it is 0.6 per thousand per annum. Determine the overall death rate per annum from this disease for male employees in Industry B standardized with respect to the age distribution of the male employees in Industry A.

17 Doctors' records of the smoking habits and ages at death of adult male patients are summarized in Table 8.41.

Table 8.41

Type of smoker	% under 40 years old	Death rate (per '000) Under 40	Death rate (per '000) 40+
Non-smokers	50	5	16
Cigarette smokers	60	7	46
Pipe smokers	40	14	28

(a) Calculate the crude death rates for the cigarette smokers and for the pipe smokers.
(b) Using the age distribution of the non-smokers as the standard, calculate the standardized death rates for the cigarette smokers and for the pipe smokers.
(c) Explain why the standardized death rates are preferable to the crude death rates for comparing the effects of cigarette smoking and pipe smoking.

18 A gardener purchased two packets of bean seeds and graded all the seeds according to size before sowing them. The number of seeds of each size in each packet and the number that germinated are shown in Table 8.42.
(a) Calculate the overall crude germination rate for each packet.
(b) Calculate the germination rate for Packet 2 standardized with respect to the size distribution of the seeds in Packet 1. Comment on what your result shows.
(c) Based on the combined results from both packets, which size of seed has the higher germination rate?

Table 8.42

Packet no.	1	1	1	2	2	2
Size	A	B	C	A	B	C
No. of seeds	20	18	12	24	12	20
No. that germinated	15	12	9	18	8	16

9
PAIRED VARIABLES

"What you see depends on
which window you look through."
Anon.

9.1 Introduction

In many statistical investigations data are collected on two or more variables. One purpose of such an investigation may be to establish whether or not there is some association between the observed values of the variables. The variables are often characteristics of individuals (or items) in a population. For instance, in Example 1.2, p. 2 data was collected on several variables for each of 20 boys and 16 girls in a class. Here we shall restrict consideration to a study of two variables only.

9.2 Scatter Diagrams

Having collected data on two quantitative variables, it is always instructive as an initial step to plot the pairs of observed values on a sheet of graph paper; the resulting diagram is referred to as a **scatter diagram**.

○ *EXAMPLE 9.1*

Table 9.1 shows the marks out of 100 obtained by 10 pupils in the two papers of a Mathematics examination.

The pairs of marks obtained by the 10 pupils have been plotted in the scatter diagram displayed in Figure 9.1, in which the horizontal axis is used for the Paper 1 marks and the vertical axis is used for the Paper 2 marks. Both axes are scaled so that 1 cm represents 5 marks.

Figure 9.1 *Scatter Diagram for the Data in Table 9.1*

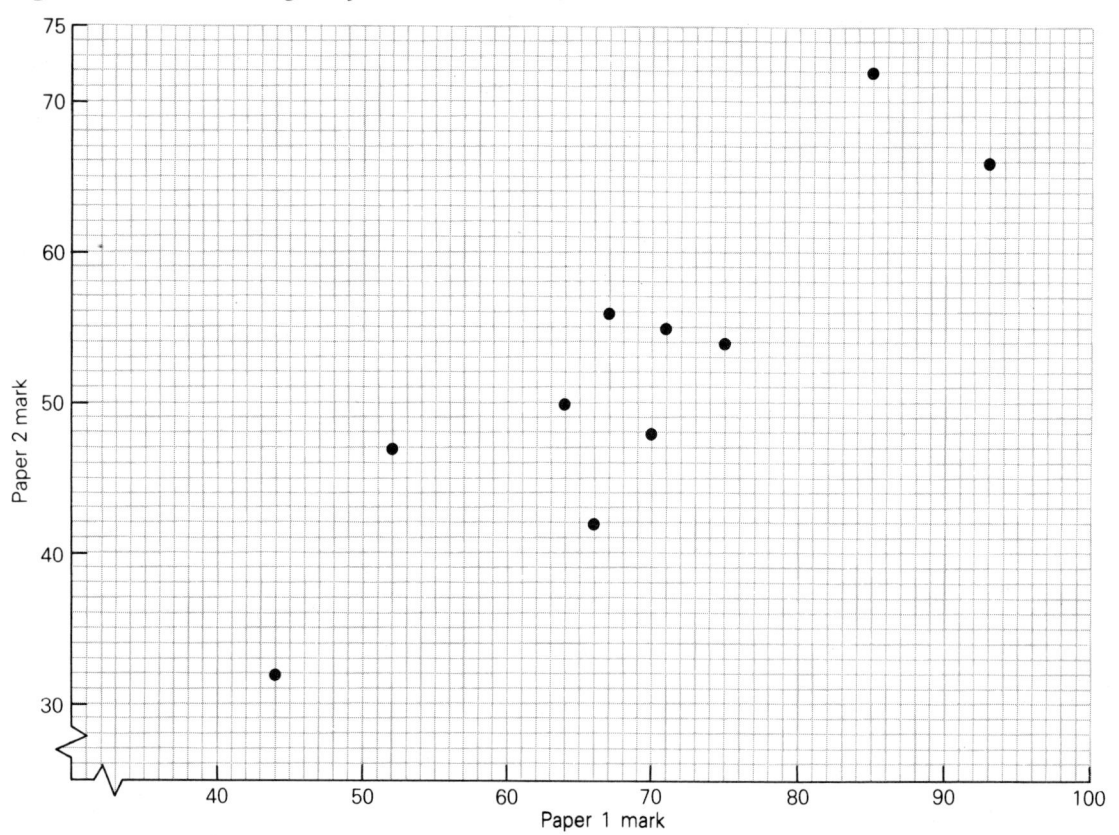

Table 9.1 *Marks Obtained by 10 Pupils in Two Mathematics Papers*

Pupil	A	B	C	D	E	F	G	H	I	J
Paper 1	71	67	52	64	44	93	66	85	75	70
Paper 2	55	56	47	50	32	66	42	72	54	48

The zig-zags at the bottom of the vertical axis and at the extreme left of the horizontal axis are there to remind you that parts of the axes are missing, as there are no zero values in this case.

In view of the use that will be made later of a scatter diagram, it is important that the scales chosen are easy to read, particularly for values intermediate between those shown along the axes.

It is clear from looking at Figure 9.1 that there is some association between the two variables (Paper 1 mark and Paper 2 mark) in that a high (low) mark in either paper is associated with a high (low) mark in the other paper. It is also apparent from the scatter of the plotted points that there is no unique relationship connecting the pairs of marks. For example, given a pupil's mark on one paper only it is not possible to calculate the precise mark the pupil obtained on the other paper.

It is important to note that the type of association we are concerned with here is not a **causal** one. In the context of the present example, a high mark in either paper does not *account* for a high mark in the other paper. The performance of a pupil on either paper is actually a measure of the pupil's mathematical ability. A pupil who is very able in Mathematics is expected to obtain high marks in both papers, while a less able pupil is expected to obtain lower marks in both papers. Contrast this with a causal association that may arise. For example, it is well known that heating a metal rod will increase its length, so that increasing the amount of heat applied to a metal rod *causes* the rod to extend in length. No such interpretation is possible in the exam paper example; the mark obtained by a pupil in one paper is dependent upon the pupil's ability and not on the mark the pupil obtained in the other paper.

○ **EXAMPLE 9.2**

Table 9.2, compiled from Table 1.2, shows the shoe sizes and the heights (to the nearest centimetre) of 20 boys in a class.

Table 9.2 *Shoe Sizes and Heights of 20 Boys*

Shoe size	5	$5\frac{1}{2}$	6	$6\frac{1}{2}$	7	$7\frac{1}{2}$	8
Height (cm)	159	167	166	170	172	174	173
	162	162	166	170	170	171	175
	163	165		168	171	172	
						173	

The scatter diagram for this data is shown in Figure 9.2. Shoe size is represented along the horizontal axis using a scale of 2 cm to each ½ shoe size, and height is represented along the vertical axis using a scale of 1 cm to each 5 cm of height. As was the case in Example 9.1, it is apparent from Figure 9.2 that there is some association between the two variables (height and shoe size) in that taller boys tend to wear larger shoes and shorter boys tend to wear smaller shoes. It is equally clear that there is no

Figure 9.2 *Scatter Diagram for the Data in Table 9.2*

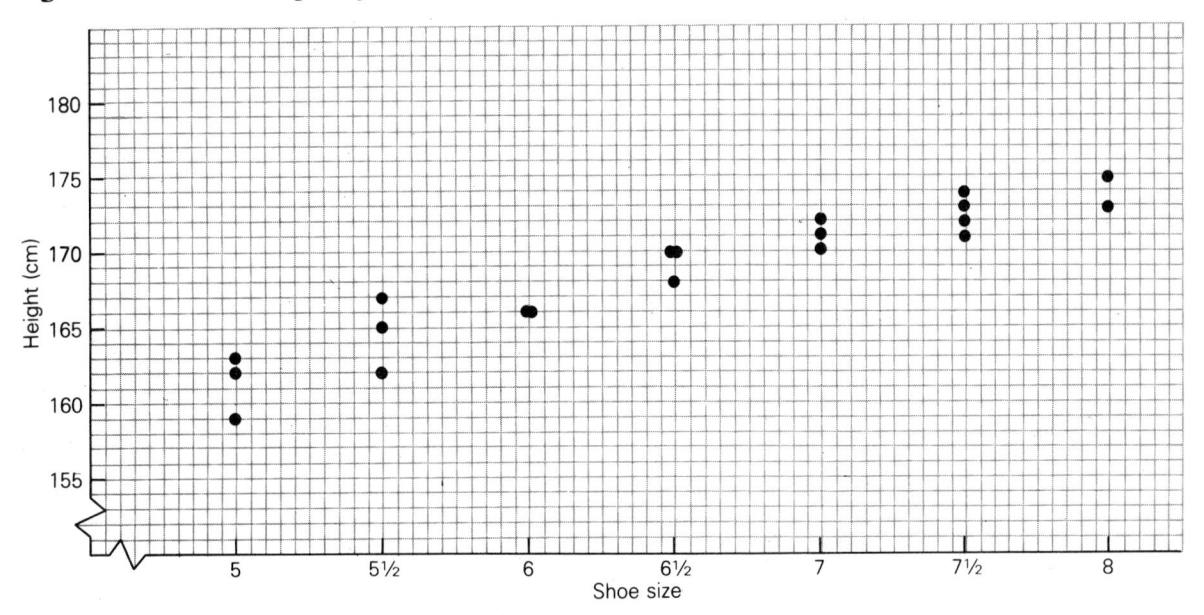

Figure 9.3 *Scatter Diagram for the Data in Table 9.3*

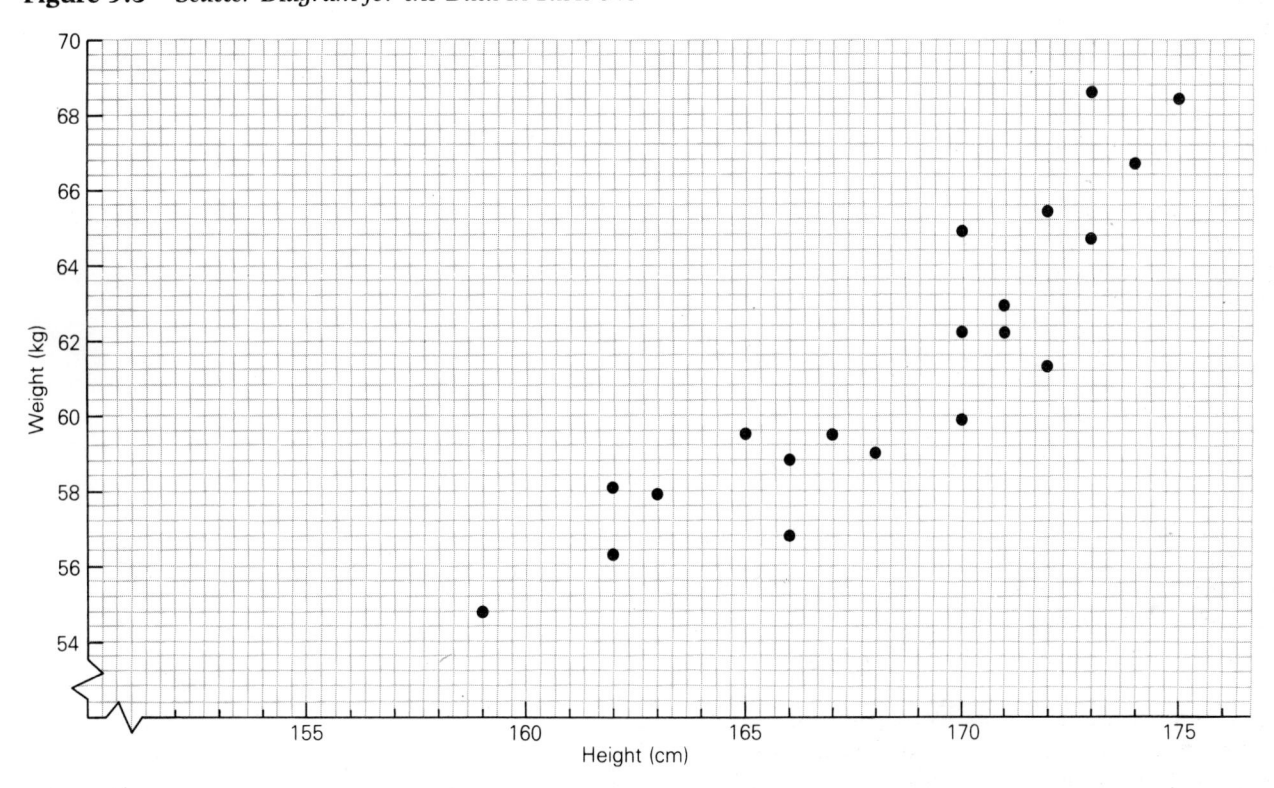

exact relationship connecting the two variables, since we find that boys of the same height may wear shoes of different sizes and boys wearing the same size of shoe may have different heights. Here again the association is not a causal one since the size of shoe that a boy will wear depends on many factors other than his height.

○ EXAMPLE 9.3

Table 9.3, compiled from Table 1.2, p. 3, shows the heights (to the nearest cm) and the weights (to the nearest 0.1 kg) of 20 boys in a class. The corresponding scatter diagram is displayed in Figure 9.3.

Here again we see that there is no unique relationship between the two variables, height and weight. But there is a distinct pattern displayed, in that large (small) values of either variable tend to be associated with large (small) values of the other variable.

○ EXAMPLE 9.4

Table 9.4 gives the heights of 10 pupils and the marks out of 100 they obtained for an English essay. The scatter diagram for this data is displayed in Figure 9.4, where the marks are represented along the horizontal axis using a scale of 1 cm to 5 marks, and the heights are represented along the vertical axis using a scale of 1 cm to 5 cm of height. Unlike the previous examples, there appears to be no trend pattern in the plotted points. This is not really surprising since it is inconceivable that there should be any association between a pupil's height and his/her ability in essay writing.

In each of Examples 9.1–9.3 the general pattern displayed by the plotted points was that one variable tended to increase in value as the value of the other variable increased. In the following example this trend is reversed in that the general pattern is for one

Table 9.3 *Heights and Weights of 20 Boys*

Height	Weight	Height	Weight
159	54·8	172	65.4
167	59.5	168	59.0
170	62.2	165	59.5
170	59.9	166	56.8
174	66.7	163	57.9
173	64.7	170	64.9
162	56.3	172	61.3
162	58.1	173	68.6
171	62.2	175	68.4
166	58.8	171	62.9

Table 9.4 *Heights and Essay marks of 10 Pupils*

Pupil	Height (cm)	Mark
A	170	71
B	167	67
C	166	52
D	159	64
E	162	44
F	167	93
G	166	66
H	164	85
I	162	75
J	164	70

Table 9.4 *Heights and Essay Marks of 10 Pupils*

variable to *decrease* in value as the value of the other variable increases.

○ EXAMPLE 9.5

Table 9.5 gives the engine sizes (cylinder volume in cubic centimetres) and the manufacturers' quoted fuel consumptions (in kilometres per litre) under urban driving conditions of 7 different car models.

Table 9.5 *Engine Sizes and Fuel Consumptions of 7 Cars*

Engine size (cc)	Consumption (km per litre)
1200	14.9
1400	12.7
1500	11.1
1600	9.9
1800	8.1
2000	7.6
2200	7.2

The scatter diagram for this data is displayed in Figure 9.5, in which the horizontal axis represents engine size using a scale of 1 cm to 100 cc, and the vertical axis represents fuel consumption using a scale of 1 cm to 2 km per litre. It is clear from this diagram that engine size and fuel consumption are associated in such a way that, as the engine size increases, the fuel consumption decreases. (But note that there are factors other than engine size that may affect a car's fuel consumption.)

● EXERCISE 9.1

Draw a scatter diagram on graph paper for each of the following sets of paired data on two variables.

Figure 9.5 *Scatter Diagram for the Data in Table 9.5*

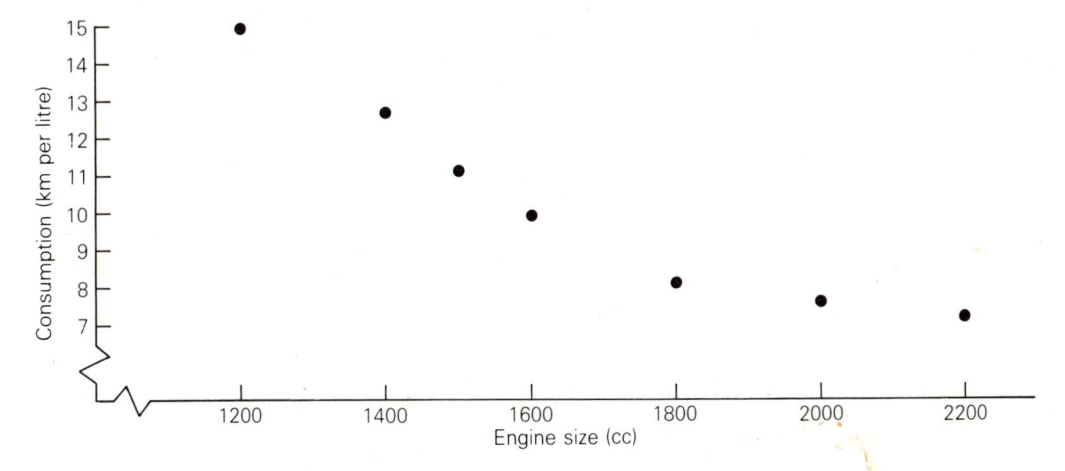

In each case state the scales you have used along the axes and the form of any general pattern displayed by the plotted points. Keep your diagrams of Questions 1–5 for use in later exercises.

1 The shoe sizes and the heights of the 16 girls as given in Table 1.2, p. 3.
2 The heights and the weights of the 16 girls as given in Table 1.2, p. 3.
3 The temperatures and lengths of a metal rod given in Table 9.6

Table 9.6 *Length of Metal Rod when Heated*

Temperature (°C)	Length of rod (m)
20	10.0
40	10.2
60	10.3
80	10.6
100	10.7

4 The ages and weights of eight babies given in Table 9.7.

Table 9.7

Age (months)	Weight (kg)
1	4.5
2	6.4
2	5.9
2	6.4
3	6.7
3	6.8
3	7.2
4	7.6

5 The paired values of two variables x and y given in Table 9.8.

Table 9.8

x	5	10	15	20	25
y	55	52	50	48	45

6 The paired values of two variables x and y given in Table 9.9.

Table 9.9

x	y
7.5	36
9	16
15	10
10	22
6	12
7	14
13	8
5	4
8	24
14.5	15
10.5	40

9.3 Line Fitting

In each of the scatter diagrams for Examples 9.1–9.3 the plotted points show a clear general pattern in that the two variables tend to increase together. It would seem reasonable in each case to approximate the trend by means of a straight line. In the case of the scatter diagram for Example 9.4 no general pattern is apparent, while in the scatter diagram for Example 9.5 it would appear that the general trend is best approximated by means of a curved line.

Here we will only consider scatter diagrams in which the general trend is approximately linear. The aim, then, is to draw a straight line on the scatter diagram which seems to be a reasonably good 'fit' to the plotted points. (A perfect fit, when all the plotted points fall exactly on a straight line, is very unlikely to arise with the sorts of variables that we are interested in.) Denote the variable which is represented along the horizontal axis by x, and the variable represented along the vertical axis by y. It is generally agreed that the fitted straight line should always pass through the point $(\bar{x}, \bar{y})$, where $\bar{x}$ and $\bar{y}$ are the means of the observed x-values and the

Figure 9.6 *Line Fitted to the Data in Table 9.1*

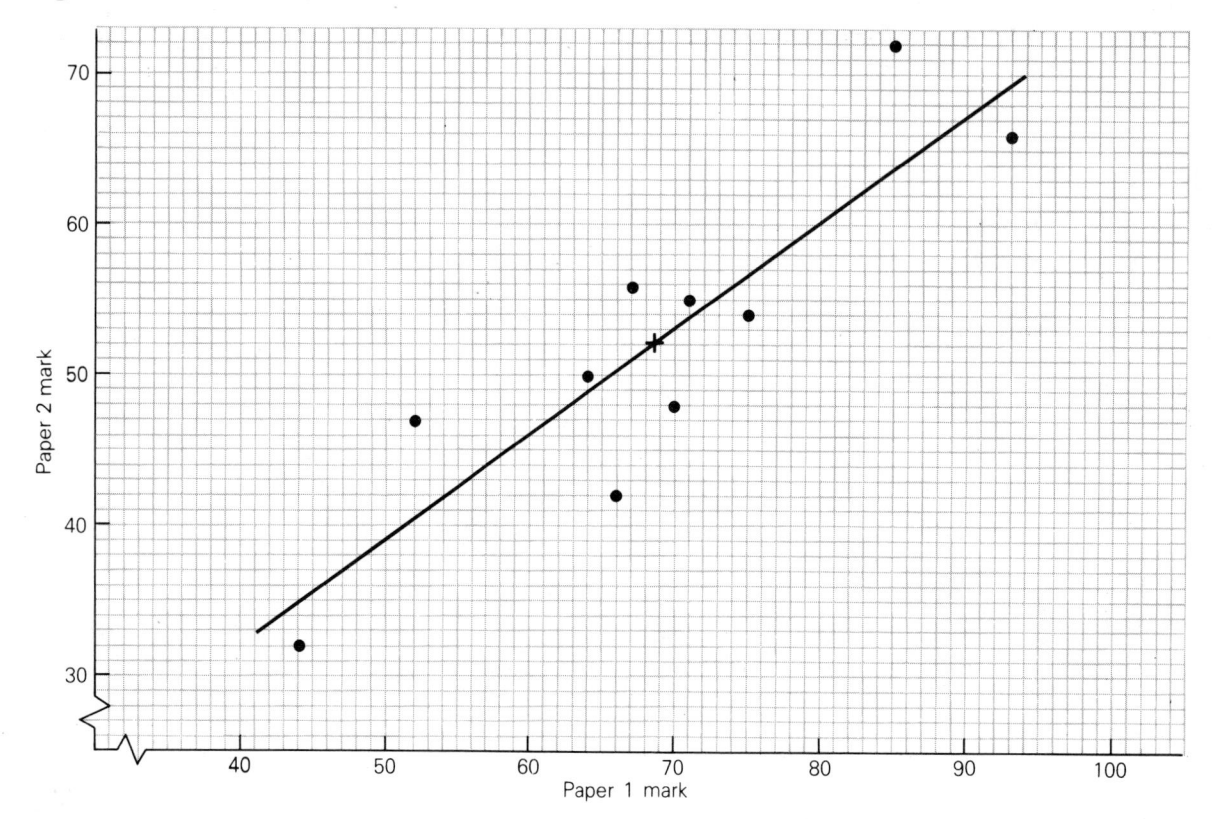

observed y-values respectively. (The justification for this is beyond the level of this book.) The gradient (slope) of the fitted line is then judged by eye; there should be about as many plotted points above the straight line as below it. A transparent ruler is very useful for this purpose.

○ EXAMPLE 9.1 (Continued)

In Example 9.1 the x-variable is the Paper 1 mark and the y-variable is the Paper 2 mark. The mean, $\bar{x}$, of the 10 pupils' marks on Paper 1 is:

$$\bar{x} = \frac{71 + 67 \ldots + 75 + 70}{10} = \frac{687}{10} = 68.7,$$

and the mean, $\bar{y}$, of their Paper 2 marks is:

$$\bar{y} = \frac{55 + 56 + \ldots + 54 + 48}{10} = \frac{522}{10} = 52.2.$$

In Figure 9.6 the data of Example 9.1 have been replotted, and the location of $(\bar{x}, \bar{y})$, that is (68.7, 52.2), has been indicated by means of a cross (**X**). Also a straight line has been drawn through (68.7, 52.2) to fit, by eye, the plotted points. Because of the fairly high degree of scatter in the plotted points it is likely that the choice of line 'fitting' them may vary slightly from person to person.

For any specified Paper 1 mark x (in the observed range from 44 to 93) the corresponding value of y given by the fitted straight line may be regarded as representative of the Paper 2 marks of pupils who obtained a mark of x on Paper 1. For example, reading from the graph in Figure 9.6 we see that for pupils obtaining 60 marks on Paper 1, a representative mark on Paper 2 is 46. Also, for pupils obtaining 75 marks on Paper 1, a representative mark for Paper 2 is about 56.7, or 57 to the nearest whole number.

Similarly, for any specified mark y on Paper 2 the corresponding value of x given by the fitted straight line is representative of the Paper 1 marks of pupils who obtained y marks on Paper 2. For example, reading from the graph in Figure 9.6 we see that for pupils obtaining 40 marks on Paper 2, a representative mark for Paper 1 is about 51.7, or 52 to the nearest whole number. Also, for pupils obtaining 65 marks on Paper 2, a representative mark for Paper 1 is 87 to the nearest whole number.

○ EXAMPLE 9.2 (Continued)

In Example 9.2 the x-variable is shoe size and the y-variable is height. Their means are:

$$\bar{x} = \frac{(5 \times 3) + (5\tfrac{1}{2} \times 3) + \ldots + (7\tfrac{1}{2} \times 4) + (8 \times 2)}{20} = 6.5,$$

and

$$\bar{y} = \frac{159 + 162 + \ldots + 173 + 175}{20} = 168.45.$$

In Figure 9.7 we have replotted the data of Example 9.2 and indicated the location of $(\bar{x}, \bar{y})$ at (6.5, 168.45) by means of a cross. A straight line has been drawn through (6.5, 168.45) which seems a reasonable fit to the plotted points. For any specified shoe size x (in the observed range) the corresponding value of y given by the fitted line is representative of the heights of boys wearing that size of shoe. For example, we can see from the graph in Figure 9.7 that a single value representation of the heights of boys wearing size 6 shoes is 167 cm. Since shoe size

Figure 9.7 *Line Fitted to the Data in Table 9.2*

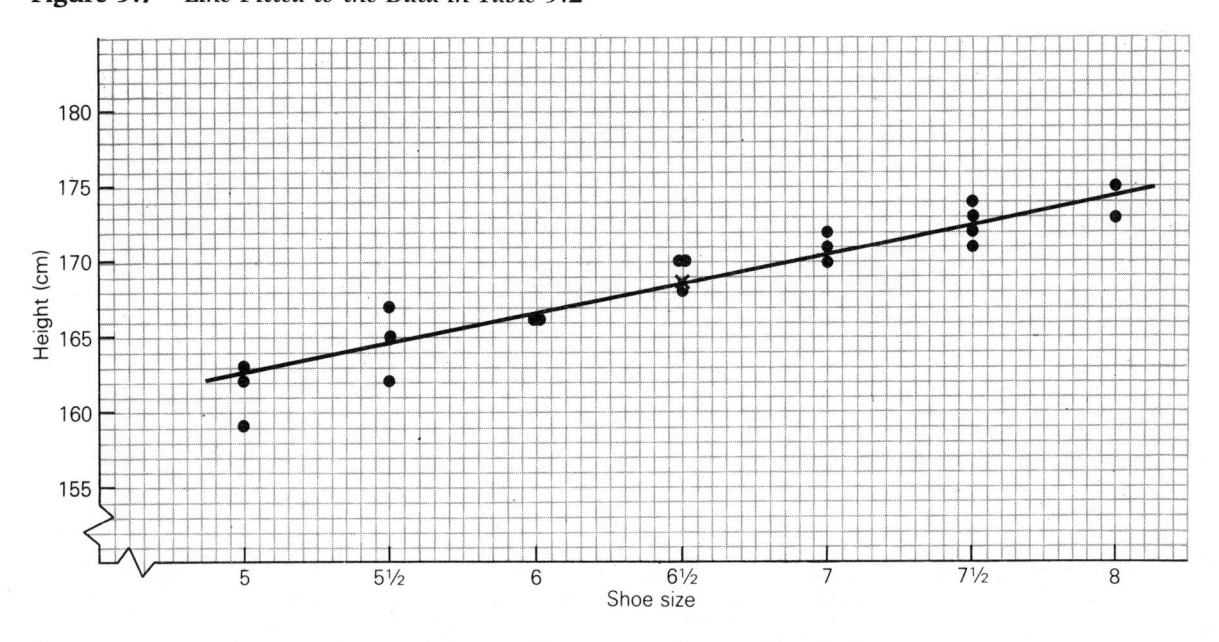

is a discrete variable it does not make sense in this example to use the fitted line to obtain a representative shoe size for boys of a given height. This contrasts with Example 9.1 in which both variables (marks on Paper 1 and Paper 2) may be regarded as being continuous so that the fitted line can be used sensibly to obtain a representative value of y for any given x and a representative value of x for any given y.

○ **EXAMPLE 9.3**   *(Continued)*

In the scatter diagram in Figure 9.3 we chose to have height as the x-variable and weight as the y-variable. As shown above the mean height of the 20 boys is:

$$\bar{x} = 168.45 \text{ cm}.$$

The mean weight of the 20 boys (from Table 9.3) is:

$$\bar{y} = \frac{54.8 + 59.5 + \ldots + 68.4 + 62.9}{20} = 61.395 \text{ kg}.$$

Figure 9.8 shows a plot of the data in Table 9.3. The location of $(\bar{x}, \bar{y})$, at (168.45, 61.395) is shown by means of a cross, and a straight line has been fitted 'by eye' through the point $(\bar{x}, \bar{y})$. Here again, because of the high degree of scatter of the plotted points the choice of line to fit the points is a matter of opinion, so different people may fit lines of different slopes through $(\bar{x}, \bar{y})$,

Using the line we have fitted in Figure 9.8 we see,

Figure 9.8 *Line Fitted to the Data in Table 9.3*

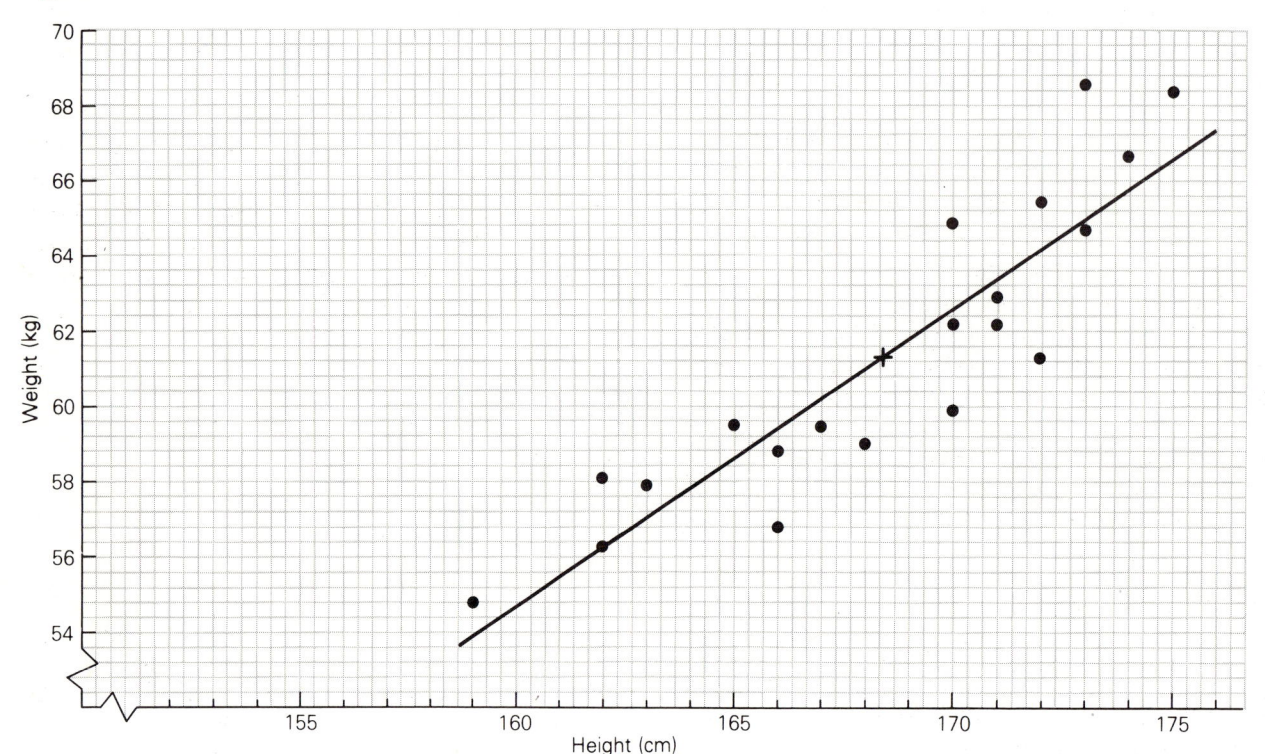

for example, that a representative value for the weights of boys having height 166 cm is 59.4 kg, and a representative value for the heights of boys weighing 64 kg is 171.6 cm.

○ **EXAMPLE 9.4**   *(Continued)*

Since no pattern is apparent in the plotted points displayed in Figure 9.4 fitting a line will serve no useful purpose. (A fitted line in this case would be virtually horizontal, indicative of no association between a pupil's height and the mark he/she obtains for an essay.)

○ **EXAMPLE 9.5**   *(Continued)*

The scatter diagram in Figure 9.5 shows that there is an association between engine size and fuel consumption, and that the trend shown would best be fitted by means of a curved line. The type of association here is different from that in previous examples since one variable tends to decrease as the other increases.

It is apparent from Figure 9.5 that for engine sizes up to 1800 cc a straight line fit to the plotted points would be very reasonable. Restricting consideration to engine sizes up to 1800 cc, the mean engine size of the five cars is:

$$\bar{x} = \frac{1200 + 1400 + 1500 + 1600 + 1800}{5} = 1500 \text{ cc},$$

Figure 9.9 *Line Fitted to the Data in Table 9.5*

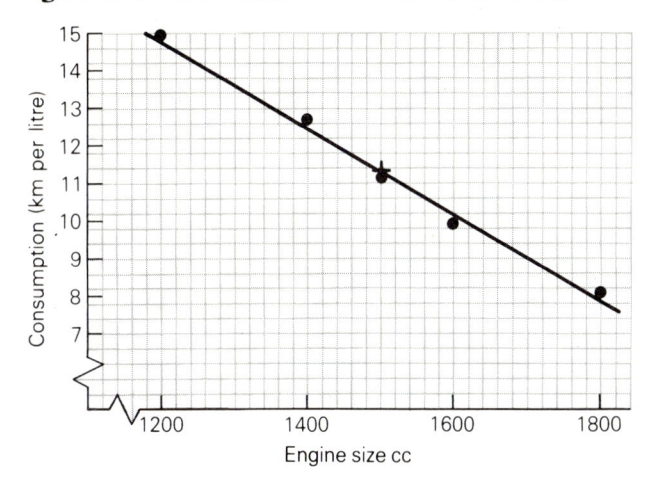

and the mean fuel consumption is:

$$\bar{y} = \frac{14.9 + 12.7 + 11.1 + 9.9 + 8.1}{5} = 11.34$$

km per litre.

A plot of the values of the engine sizes and fuel consumptions of the five cars being considered is shown in Figure 9.9. The location of $(\bar{x}, \bar{y})$ at (1500, 11.34) is indicated by a cross, and a straight line has been fitted through that point by eye. In this particular example the band of points is rather narrow, so that lines fitted by different people should not be very different.

Using the fitted line we find, for example, that a representative fuel consumption for cars of engine size 1700 cc is about 9.0 km per litre. In this example, since engine sizes are fairly standard, there is little point in estimating the engine size that will give a specified fuel consumption.

A word of caution is in order at this juncture. Great care is needed if a fitted line is to be extended beyond the ranges of the observed values of the variables for the purpose of estimation. This is referred to as **extrapolation** and it should be done only when there is ample evidence to justify the assumption that the trend displayed in the scatter diagram may be extended. In the present example it is evident that the linear trend shown by the plotted points in Figure 9.9 does not extend to cars of engine sizes greater than 1800 cc. (Refer back to Table 9.5 or Figure 9.5 for the fuel consumptions of the cars of engine sizes 2000 cc and 2200 cc.)

● **EXERCISE 9.2**

For each of the scatter diagrams you drew in Questions 1–5, Exercise 9.1, fit a straight line, through $(\bar{x}, \bar{y})$, to the plotted points. Keep your graphs for use in Exercise 9.3.

1 Use your fitted line on the scatter diagram for Question 1, Exercise 9.1 to estimate a representative height of girls who wear shoes of size (a) $4\frac{1}{2}$ (b) 6.

2 Use your fitted line on the scatter diagram for Question 2, Exercise 9.1 to estimate (a) a representative weight for girls of height (i) 145 cm (ii) 160 cm (b) a representative height for girls weighing (i) 50 kg (ii) 60 kg.

3 Use your fitted line on the scatter diagram for Question 3, Exercise 9.1 to estimate (a) a representative length of the rod when it is heated to a temperature of (i) 30°C (ii) 75°C (b) a representative temperature for the rod to have length (i) 10.25 m (ii) 10.5 m.

4 Use your fitted line in Question 4, Exercise 9.1 to estimate a representative weight of babies aged (a) $2\frac{1}{2}$ months, (b) 3 months.

5 Use your fitted line in Question 5, Exercise 9.1 to estimate (a) a representative value for y when (i) x=3 (ii) x=21 (b) a representative value for x when (i) y=51 (ii) y=46.

9.4 Slope of a Fitted Line

A general form for the equation of a straight line is:

$$y = mx + c.$$

This means that the x-value and the y-value of any point on a straight line will satisfy this equation for particular values of m and c. Observe that c is the value of y when x=0, and it is the y-value of the **intercept** (intersection) of the straight line with the y-axis. This is illustrated in Figure 9.10.

Figure 9.10 *Intercept of* $y = mx + c$: *(a) Positive c, (b) Negative c*

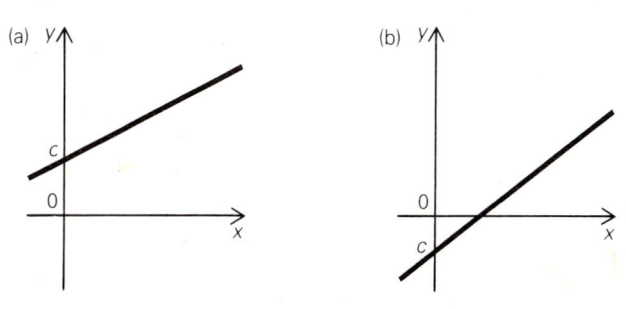

In Figure 9.10a the value of c is positive, and in Figure 9.10b its value is negative. The constant m is the **slope** (or **gradient**) of the straight line and is the amount by which the value of y will change if the value of x is increased by one unit. If the line slopes upwards from left to right then m is positive, and the value of y will *increase* by m units when x is increased by one unit (see Figure 9.11a). If the line slopes downwards from left to right then m is negative, and the value of y will *decrease* by m units when the value of x increases by one unit (see Figure 9.11b). (For readers who are familiar with trigonometry, the slope m is the tangent of the angle that the straight line makes with the *positive* part of the x-axis.)

Figure 9.11 *Slope of* $y = mx + c$: *(a) Positive m, (b) Negative m*

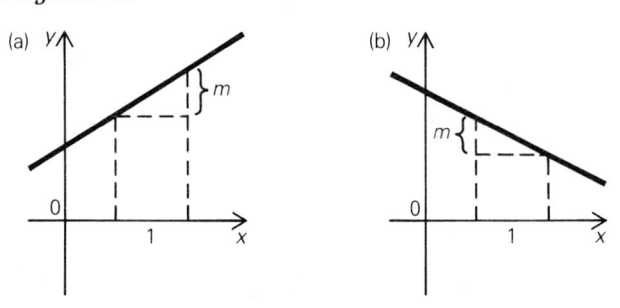

The lines that we fitted in Section 9.3 always passed through the fixed point $(\bar{x}, \bar{y})$. For such a line the values of m and c must be such that:

$$\bar{y} = m\bar{x} + c.$$

Line Fitted to the Data in Table 9.2

Hence, $c = \bar{y} - m\bar{x}$, and the equation of the fitted line can be written as:

$$y = mx + \bar{y} - m\bar{x},$$

or equivalently,

$$y = \bar{y} + m(x - \bar{x}), \qquad (9.1)$$

which is a more convenient form for our purposes. Now consider how we can determine the value of m for a particular line fitted to a scatter diagram. To do so, all we need do is find the change in the value of y when x is increased by one unit. A graphical method for achieving this is illustrated in the following examples.

○ **EXAMPLE 9.2** *(Continued)*

Determine the equation of the fitted line in Figure 9.7.

For convenience Figure 9.7 is reproduced here. Consider the change in the value of y when x is increased from 6 to 7. From the graph we see that when $x = 7$ the value of y is approximately 170.8, and when $x = 6$ the value of y is approximately 166.5. Hence for a unit increase in x (from 6 to 7) the value of y has increased by $170.8 - 166.5 = 4.3$, approximately. Therefore the slope of the fitted line is approximately $m = 4.3$.

We showed earlier that for this example $\bar{x} = 6.5$ and $\bar{y} = 168.45$. It follows that the equation of our fitted line, using Equation (9.1), is approximately:

$$y = 168.45 + 4.3(x - 6.5),$$

or:

$$y = 4.3x + 140.5.$$

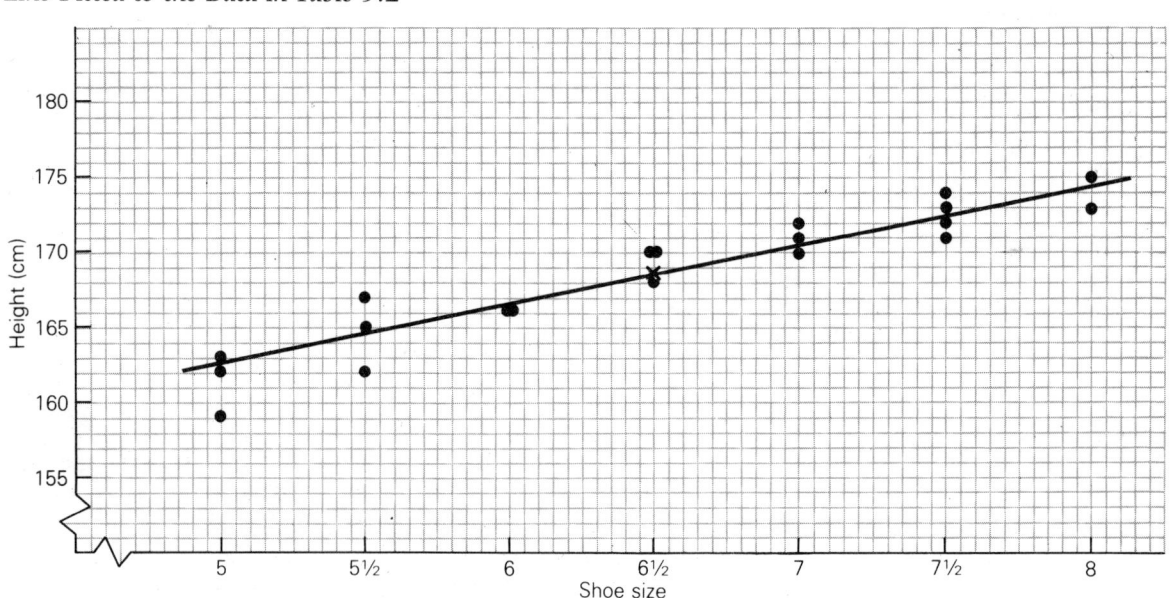

Observe that the intercept of the line (that is, the value of y when $x=0$) is $c=140.5$, but this has no meaningful practical interpretation in this example since it is absurd to suggest that for boys of height 140.5 cm the representative shoe size is zero! This is another example where extrapolation is unwarranted.

We could now use Equation (9.1) instead of the graph to estimate a representative value of y corresponding to any given shoe size x. For example, from the equation a representative height for boys wearing shoes of size 6 is given by:

$$y = (4.3 \times 6) + 140.5 = 166.3 \text{ cm}.$$

Compare this with the estimate of 167 cm obtained earlier directly from the graph. Slight discrepancies between estimates obtained graphically and by calculation using this method are inevitable because of errors in reading from the graph.

When determining the slope of a fitted line from its graph it is usually more accurate to find the change in the value of y for two values of x that are far apart, preferably near the extremes of the observed values. Dividing this change in the value of y by the difference between the two chosen x-values will then give the value of the slope m. Using the fitted line in Figure 9.7 as an example, let us find the values of y when $x = 5$ and $x = 8$. From the fitted line we find that when $x = 8$ the value of y on the line is approximately 174.8, and when $x = 5$ the value of y on the line is approximately 162.5. Using these results the slope of the fitted line is approximately.

$$m = \frac{174.8 - 162.5}{8 - 5} = 4.1.$$

This differs slightly from our earlier estimate, but is probably more accurate because the effect of any slight errors in reading y-values from the graph will be lessened due to division by the difference between the two x-values used (a number greater than 1).

Thus a more accurate approximation to the equation of the line fitted in Figure 9.7 is:

$$y = 168.45 + 4.1(x - 6.5),$$

or

$$y = 4.1x + 141.8$$

When $x = 6$ this gives $y = 166.4$ cm as a representative height of boys wearing size 6 shoes.

○ **EXAMPLE 9.5** (Continued)

Determine the equation of the fitted line in Figure 9.9.

To determine the slope of the fitted line consider the change in the value of y (fuel consumption) when x (engine size) is increased from 1200 cc to 1800 cc (the two extreme observed values). From Figure 9.9 we find that when $x = 1200$ the value of y on the line is approximately 14.8, and when $x = 1800$ the value of y on the line is approximately 7.8. Thus, corresponding to an increase in x of $1800 - 1200 = 600$ units, the value of y has *decreased* by $14.8 - 7.8 = 7$ units. Since y has decreased, the slope m is negative and is approximately:

$$m = -\frac{7}{600} = -0.0117.$$

Having shown earlier that $\bar{x} = 1470$ and $\bar{y} = 11.67$ it follows, using Equation (9.1), that the equation of our fitted line is approximately:

$$y = 11.67 - 0.0117(x - 1470),$$

or equivalently:

$$y = 28.87 - 0.0117x.$$

As in the previous example we could now use this equation to estimate a representative value for the fuel consumption y of cars having any specified engine size x.

● **EXERCISE 9.3**

1 Determine the equation of the fitted line in Figure 9.6 and use it to estimate:
(a) a representative Paper 2 mark for a Paper 1 mark of (i) 65 (ii) 88,
(b) a representative Paper 1 mark for a Paper 2 mark of (i) 40 (ii) 60.
2 Determine the equation of the fitted line in Figure 9.8 and use it to estimate:
(a) a representative weight of boys of height (i) 163 cm (ii) 171 cm,
(b) a representative height of boys weighing (i) 58 kg, (ii) 65 kg.
Questions *3–7* For each of Questions 1–5, Exercise 9.2 find the equation of your fitted line and use it to obtain the estimates asked for in Exercise 9.2.

9.5 The Line of 'Best' Fit*

We will now consider a method of fitting a straight line to the points in a scatter diagram to estimate the mean value of y, as opposed to an arbitrary representative value as in Section 9.4. For a given value of x it may be shown (using a principle known as 'least squares' which will not be described here) that the 'best' line is the one through $(\bar{x}, \bar{y})$ having slope m, given by the formula:

$$m = \frac{n\Sigma xy - (\Sigma x)(\Sigma y)}{n\Sigma x^2 - (\Sigma x)^2}. \quad (9.2)$$

Here n is the number of observed values (x, y), Σx and Σy are the sums of the n values of x and the n values of y respectively, and Σxy is the sum of the n products of corresponding pairs of values of x and y. The resulting line will have the equation:

$$y = \bar{y} + m(x - \bar{x}), \quad (9.3)$$

with m as given above, and is referred to as the **least squares regression line of y on x**. (In most examinations the equation of this line and the formula for m will be in the information booklet or stated on the examination paper.)

Consider Example 9.2 again in which we had observations on the shoe sizes, x, and heights, y, of 20 boys. Suppose we want to estimate the mean height of boys wearing a specified size of shoe. From the data in Table 9.2 we find:

$\Sigma x = (5 \times 3) + \ldots + (8 \times 2) = 130$,

$\Sigma x^2 = 5^2 \times 3 + \ldots + 8^2 \times 2 = 864.5$,

$\Sigma y = 159 + 162 + \ldots + 173 + 175 = 3369$,

$\Sigma xy = 5(159 + 162 + 163) + \ldots + 8(173 + 175)$
$= 21981$.

From Expression (9.2) the slope of the least squares regression line of y on x is:

$$m = \frac{(20 \times 21\,981) - (130 \times 3369)}{(20 \times 864.5) - 130^2} = \frac{1650}{390} = 4.231,$$

to three decimal places. (Compare this with the slope of the line we fitted by eye in Figure 9.7 which was 4.1.) Substituting in Equation (9.3), the equation of the least squares regression line of height y on shoe size x is:

$$y = \frac{3369}{20} + 4.231\left(x - \frac{130}{20}\right),$$

or

$$y = 4.231x + 140.95. \quad (9.4)$$

Using Equation (9.4) the estimate of the mean height of boys wearing shoes of size 6 is $(4.231 \times 6) + 140.95 = 166.3$ cm. Compare this with the value 166.4 cm we obtained earlier from the line fitted by eye.

Now consider Example 9.5, restricted to cars with engine sizes from 1200 cc to 1800 cc inclusive. Here the number of observed values is $n = 5$, x denotes engine size and y denotes fuel consumption. We find from Table 9.5 that:

$\Sigma x = 1200 + 1400 + 1500 + 1600 + 1800 = 7500$,
$\Sigma x^2 = 1200^2 + \ldots + 1800^2 = 11\,450\,000$,
$\Sigma y = 14.9 + 12.7 + 11.1 + 9.9 + 8.1 = 56.7$,
$\Sigma xy = (1200 \times 14.9) + \ldots + (1800 \times 8.1) = 82\,730$.

From Expression (9.2) the slope of the least squares regression line of y on x is given by:

$$m = \frac{(5 \times 82\,730) - (7500 \times 56.7)}{(5 \times 11\,450\,000) - (7500)^2}$$
$$= -\frac{11\,600}{1\,000\,000} = -0.0116.$$

Compare this with our estimate of -0.0117 for the slope of the line we fitted by eye in Figure 9.9. It follows that the equation of the least squares regression line of fuel consumption y on engine size x is:

$$y = \frac{56.7}{5} - 0.0116\left(x - \frac{7500}{5}\right)$$

or

$$y = 28.74 - 0.0116x.$$

In particular, the least squares estimate of the mean fuel consumption of cars of engine size 1700 cc is $28.74 - (0.0116 \times 1700) = 9.02$ km per litre. Compare this with our earlier estimate, based on a line fitted by eye, of 9 km per litre.

● EXERCISE 9.4

1 *(a) Obtain the equation of the least squares regression line of weight on height using the data in Table 9.3, p. 106.*
(b) Use this equation to estimate the mean weight of boys of height 166 cm. (Note that in Section 9.3 we had an estimate of 59.4 kg for a representative weight of boys of height 166 cm.)
2 *(a) Obtain the equation of the least squares regression line of the length of the rod on temperature using the data given in Question 3, Exercise 9.1.*
(b) Use the equation to estimate the length of the rod when the temperature is (i) 30°C, (ii) 75°C. (Compare your answers with those you obtained in Question 3, Exercise 9.2.)
3 *(a) Determine the equation of the least squares regression line of y on x for the data given in Question 5 of Exercise 9.1. (b) Hence estimate the mean value of y when (i) x=8, (ii) x=21. (Compare your answers with those you obtained in Question 5(a), Exercise 9.2.)*

ASK A SILLY QUESTION...

Many employers use selection tests to help them choose future employees. The value of such tests will depend on how good the questions in them are for predicting people who will make good workers and who will not.

To obtain a test, we might identify people who are already in employment, doing the relevant kind of work. These people can then be ranked in terms of the quality of their work. A large number of possible questions are written and put to these people. Those questions which regularly attract one response from "good" workers and a different response from "bad" workers are assumed to be good discriminators. These are the questions that will be used as the selection instrument for future employees.

Suppose, for the sake of argument, that a question was originally included that said: "Do you own a car?" It might be found that "good" employees said 'yes', and that "less good" employees said 'no', so this question looks like a good discriminator. When put to potential employees, however, hardly any responded 'yes', certainly not enough to enable the firm to maintain an adequate workforce if they only employed the 'yes' respondents.

The problem here is one of spurious correlation. Car ownership is probably associated with salary received, which is in turn probably correlated with quality of work done or status in the original employees' ranking.

Quality of Work
↓
Salary
↓
Car Ownership

Potential employees may well be younger persons, not necessarily already earning, and therefore unlikely to own cars. The question, then, is not a good discriminator for the population of people for whom the test is intended. How often are tests used that have been derived on false or indirect correlations? What are the alternatives to tests created in this way?

9.6 Correlation and Spearman's Rank Correlation Coefficient

When the scatter diagram for pairs of observed values of two variables shows a pattern of association we say that the two variables are **correlated**. If the association is such that one of the variables tends to increase as the value of the other variable increases then we say that the variables are **positively correlated**. The scatter diagrams of Figures 9.1–9.3 show that the corresponding variables are positively correlated. If the association between the values of the variables is such that one of the variables tends to *decrease* as the other variable increases then we say that the variables are **negatively correlated**. An example is shown in Figure 9.5. Finally, if there is no apparent association between the values of two variables we say that the two variables are **uncorrelated**. An example is shown in Figure 9.4.

In the preceding sections we concentrated on a linear association between two variables, and we shall continue to do so here. It would be advantageous to have a quantitative assessment of the extent to which pairs of observed values of two variables conform with a linear pattern. One such measure is the **product moment correlation coefficient**, which will be discussed in the next section. Here we shall introduce a measure which is easier to calculate and is appropriate for variables which are measured on an ordinal (or rank) scale (see Section 1.4, p. 5).

○ EXAMPLE 9.6

Consider the marks obtained by 10 pupils in two exam papers as given in Table 9.1. Let us now rank the pupils' performances in each paper by assigning rank 1 to the pupil with the highest mark, rank 2 to the pupil with the second highest mark, and so on, ending with the pupil with the lowest mark being assigned rank 10. (The method described here is equally valid if the rank order is reversed, so that the lowest mark is assigned rank 1, the next lowest, rank 2, and so on.) For the data in Table 9.1, the ranks of the pupils on the two papers are as shown in Table 9.10.

Table 9.10 *Rankings of 10 Pupils in Two Mathematics Papers*

Pupil	A	B	C	D	E	F	G	H	I	J
Paper 1 rank	4	6	9	8	10	1	7	2	3	5
Paper 2 rank	4	3	8	6	10	2	9	1	5	7

Now consider the extent to which the ranks on the two papers agree. To do so, consider the discrepancies between the pairs of ranks. Small discrepancies will indicate good agreement and large discrepancies will indicate poor agreement. The discrepancies measured by:

$$d = \text{Paper 1 rank} - \text{Paper 2 rank},$$

are:

$$0, 3, 1, 2, 0, -1, -2, 1, -2, -2.$$

Observe that the d's should always sum to zero; this should be checked each time.

For an overall measure of the discrepancy between the pairs of marks we need to combine the d's in some convenient way. Summing them would be inappropriate because the answer will always be zero. The combination used in practice is that obtained by summing the squares of the d's. For the present example the sum of the squares of the d's is given by:

$$S = 0^2 + 3^2 + \ldots + (-2)^2 = 28.$$

To assess the significance of this value of S we need to know the smallest and largest possible values that S may take and the conditions under which these values are attained. Since each contributing term to the value of S must be non-negative (each being the square of a number), it follows that the smallest possible value of S is zero. This can only arise if each $d = 0$, corresponding to the pairs of ranks being identical. For the largest possible value of S we need to look at the worst possible configuration that could arise. The worst that can occur is that one set of ranks is the complete reverse of the other set of ranks. That is, the pupil ranked 1 on Paper 1 is ranked 10 on Paper 2, the pupil ranked 2 on Paper 1 is ranked 9 on Paper 2, and so on, as shown in Table 9.11.

Table 9.11 *Worst Possible Configuration*

Paper 1 rank	1	2	3	4	5	6	7	8	9	10
Paper 2 rank	10	9	8	7	6	5	4	3	2	1

In this case we find that the value of S, its largest possible value, is given by:

$$S = (1-10)^2 + (2-9)^2 + \ldots + (10-1)^2$$
$$= 9^2 + 7^2 + \ldots + 7^2 + 9^2 = 330.$$

Thus the possible values of S may range from 0 (when there is complete agreement) to 330 (when there is complete disagreement). Our calculated value $S = 28$ is much closer to 0 than to 330, which suggests that there is some degree of agreement between the two sets of ranks.

In the general case, where n items have been ranked in two ways, the smallest possible value of S is still zero and it may be shown that its largest possible value is $\frac{1}{3}n(n^2 - 1)$. Thus a crude method for determining the extent of agreement between the two sets of ranks is to assess the calculated S in relation to the values 0 and $\frac{1}{3}n(n^2 - 1)$. The closer the calculated S is to zero the stronger is the agreement between the sets of ranks.

In 1904, the British psychologist, C.E. Spearman, suggested that instead of S we should use the standardized measure r_S defined by:

$$r_S = 1 - \frac{6S}{n(n^2 - 1)},$$

which is known as **Spearman's rank correlation coefficient**. (In most examinations this formula will appear in the information booklet or be stated in the relevant question.)

When $S = 0$ we see that $r_S = 1$, and when $S = \frac{1}{3}n(n^2 - 1)$ then $r_S = -1$. Thus the smallest and largest possible values of r_S are -1 and $+1$ respectively. A calculated r_S close to $+1$ (that is, S close to 0) indicates good agreement between the sets of ranks. A calculated r_S close to -1 (that is, S close to its greatest possible value) indicates very poor agreement.

In the present example $n=10$ and $S=28$, so that:

$$r_S = 1 - \frac{6 \times 28}{10(100-1)} = 0.83 \quad \text{(to 2 d.p.)}.$$

Since this value is fairly close to $+1$ we conclude there is good agreement between the ranks of the pupils on the two papers.

○ EXAMPLE 9.7

Six bulls at an agricultural show were ranked in order of merit by the Official Judge and by each of two assistants who were training to be judges. The rankings given by the Official Judge and the two assistants are given in Table 9.12. Determine which of the two assistants assigned ranks closest to those of the official judge.

Table 9.12 *Rankings of Six Bulls*

Official Judge	1	2	3	4	5	6
Assistant A	2	1	3	5	4	6
Assistant B	1	2	5	3	6	4

The discrepancies between the rankings given by Assistant A and the Official Judge are:

$$-1, 1, 0, -1, 1, 0.$$

Checking, these do sum to zero. The sum of the squares of these discrepancies is:

$$S_A = 1 + 1 + 0 + 1 + 1 + 0 = 4.$$

Similarly, for the discrepancies between Assistant B and the Official Judge we find that the sum of the squares is:

$$S_B = 0 + 0 + 4 + 1 + 1 + 4 = 10.$$

Since S_A is less than S_B, the rankings given by Assistant A are closer to those of the Official Judge.

Alternatively, since $n=6$ the rank correlation coefficients are:

$$r_S(A) = 1 - \frac{6 \times 4}{6 \times 35} = \frac{31}{35},$$

and:

$$r_S(B) = 1 - \frac{6 \times 10}{6 \times 35} = \frac{25}{35}.$$

Since $r_S(A)$ is greater than $r_S(B)$, the conclusion, as before, is that Assistant A's rankings are closer to those of the Official Judge.

● EXERCISE 9.5

1 A teacher ranked her eight pupils for their ability in Statistics. Her ranks and those of the marks obtained by the pupils in a test are given in Table 9.13. Calculate the value of the rank correlation coefficient between these two sets of ranks. Interpret your answer.

Table 9.13

Pupil	A	B	C	D	E	F	G	H
Teacher's ranks	3	2	1	5	8	7	6	4
Test ranks	4	1	3	6	5	8	7	2

2 Two judges at a dog show gave the ranks in Table 9.14 to the six dogs entered for a particular class. Calculate a measure of the extent of the agreement between the two judges and comment on its value.

Table 9.14

Dog	A	B	C	D	E	F
Judge 1 ranks	3	2	4	1	5	6
Judge 2 ranks	2	5	1	3	6	4

3 Two experts were asked to rank 7 antiques according to their age. The actual age ranks (from oldest to youngest) and the ranks given by the two experts are shown in Table 9.15. Determine which of the two experts is the more knowledgeable on dating antiques.

Table 9.15

Actual age ranks	1	2	3	4	5	6	7
Expert A ranks	3	1	2	6	5	4	7
Expert B ranks	2	3	1	6	4	7	5

□□ r_S AS A MEASURE OF LINEARITY

We showed earlier that $r_S = +1$ only if $S=0$, which occurs only when the two sets of ranks are identical. The scatter diagram in this case will consist of the points $(1, 1) (2, 2), (3, 3)$, and so on. Figure 9.12a is a scatter diagram for the case when $n=10$. Observe that the points fall *exactly* on a straight line which passes through the origin and has slope $m=+1$.

Now $r_S = -1$ only if S has its maximum value, which occurs only if one set of ranks is the complete reverse of the other set. The scatter diagram in this

Figure 9.12 *Extreme Cases of Two Sets of Ranks: (a) Perfect agreement, (b) Perfect disagreement*

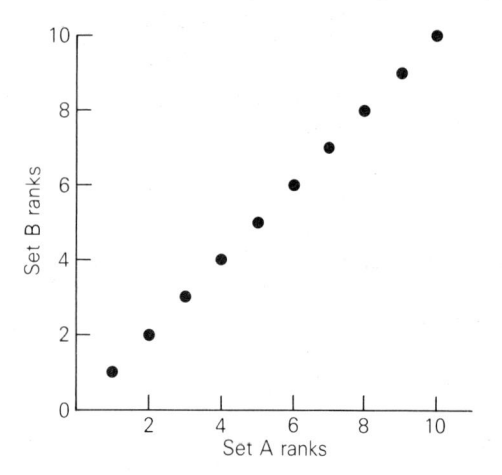

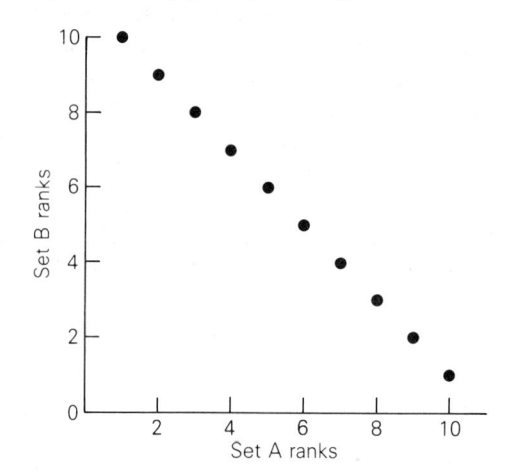

case for $n=10$ is shown in Figure 9.12b. Observe that the points fall *exactly* on a straight line through the origin which has slope $m=-1$.

Thus the value of r_S gives a measure of the extent to which pairs of ranks display a linear pattern.

(1) If r_S has a value close to $+1$ then the scatter diagram of the ranks will show a clear linear pattern having positive slope. This means that there is a high degree of agreement between the two sets of ranks.
(2) If r_S has a value close to -1 then the scatter diagram of the ranks will show a clear linear pattern with negative slope. This means that there is a high degree of disagreement between the two sets of ranks.
(3) If r_S has a value (positive or negative) close to zero then the scatter diagram of the ranks will not show any linear pattern.

In Example 9.6 our calculated r_S was 0.83, which is close enough to $+1$ to conclude that there is good agreement between the Paper 1 ranks and the Paper 2 ranks. This can be interpreted as meaning that the *marks* in the two papers have a high positive correlation, which is evidently so from Figure 9.1. When actual observed values are converted into ranks it is clear that some information is lost. For this reason, if we do convert observed values into ranks then the rank correlation coefficient should only be regarded as an *approximate* measure of the actual correlation between the observed values. This is illustrated in the following example.

○ **EXAMPLE 9.8**

Convert the data in Example 9.5 into ranks and calculate the rank correlation coefficient between engine size and fuel consumption.

From Table 9.5 the ranks of the engine sizes and fuel consumptions are as shown in Table 9.16.

Table 9.16 *Ranked Data from Table 9.5*

Engine size rank	1	2	3	4	5	6	7
Fuel consumption rank	7	6	5	4	3	2	1

Since one set of ranks is completely the reverse of the other set, $r_S = -1$. (You can check this by calculation.) This means that in the scatter diagram the plotted points fall exactly on a straight line of slope -1. Referring to the scatter diagram for the actual data in Figure 9.5, the plotted points do not fall exactly on a straight line, but they do show a strong degree of negative correlation. In this example Spearman's rank correlation coefficient has exaggerated the true situation. In general, the rank correlation coefficient is a useful *approximate* measure of the degree of linearity shown by pairs of observed values.

☐☐ THE CASE OF TIED RANKS

When actual observed values of two variables are converted into ranks it may happen that some values occur more than once. When this is so it is customary to give each repeated value the average of the ranks that they would have been given had they been slightly different. For example, suppose the observed values of one variable were:

10.2, 10.2, 10.2, 10.8, 10.8, 11.0, 11.4, 11.4.

The smallest value, 10.2, occurs three times. Had the three values been slightly different they would have been given ranks 1, 2 and 3. The average of these ranks is $(1+2+3)/3 = 2$, so we give rank 2 to each of the three observations having the value 10.2. The next smallest value is 10.8, which occurs twice. If these two observations had differed slightly they would have been given ranks 4 and 5. The average of these is $(4+5)/2 = 4\frac{1}{2}$, so each of the two observation Shaving the value 10.8 is given rank $4\frac{1}{2}$. The next observation has the value 11.0. Since this value occurs once only it is given rank 6. Finally, the largest value, 11.4, occurs twice and each of these observations is given rank $(7+8)/2 = 7\frac{1}{2}$.

○ EXAMPLE 9.9

The marks out of 10 awarded to eight television programmes by two critics are shown in Table 9.17. Rank the two sets of marks and calculate Spearman's rank correlation coefficient.

Table 9.17 *Critics' Marks for Eight Programmes*

Programme	A	B	C	D	E	F	G	H
Critic 1	6	7	6	4	7	8	9	6
Critic 2	4	9	4	8	5	9	9	4

We shall rank each set of marks from highest (rank 1) to lowest (rank 8).

For Critic 1 the highest mark is 9 for programme G, so G is assigned rank 1. The next highest mark is 8, awarded to F, so F is assigned rank 2. The next highest mark is 7, awarded to B and E, so both B and E are assigned rank $\frac{1}{2}(3+4) = 3\frac{1}{2}$. The next highest mark is 6, awarded to A, C, and H, so each of A, C and H is assigned rank $\frac{1}{3}(5+6+7) = 6$. The next mark, which is the lowest, is 4, awarded to D, so D is assigned rank 8.

The ranks for the marks awarded by Critic B are obtained similarly. The two sets of ranks are shown in Table 9.18.

Table 9.18 *Critics' Marks for Eight Programmes*

Programme	A	B	C	D	E	F	G	H
Critic 1	6	$3\frac{1}{2}$	6	8	$3\frac{1}{2}$	2	1	6
Critic 2	7	2	7	4	5	2	2	7

The discrepancies:

$$d = \text{Critic 1 rank} - \text{Critic 2 rank},$$

are:

$$-1, 1\frac{1}{2}, -1, 4, -1\frac{1}{2}, 0, -1, -1.$$

These do sum to zero. The sum of their squares is:

$$S = (-1)^2 + (1\frac{1}{2})^2 + (-1)^2 + 4^2 + (-1\frac{1}{2})^2 + 0^2 + (-1)^2 + (-1)^2 = 24\frac{1}{2}.$$

Hence, since $n = 8$, Spearman's rank correlation coefficient is equal to:

$$r_S = 1 - \frac{6 \times 24\frac{1}{2}}{8 \times 63} = 1 - \frac{3 \times 49}{8 \times 63} = \frac{17}{24} = 0.71,$$

to two decimal places. Since this value is fairly close to $+1$ we conclude that the two critics' ranks of the programmes are in good agreement. (Draw a scatter diagram of the ranks in order to see the linear pattern with a positive slope.)

☐☐ INTERPRETING A HIGH OR LOW RANK CORRELATION COEFFICIENT

In all the examples we have discussed so far the value of r_S has been fairly close to $+1$ or -1, indicating that there is a strong linear pattern displayed in the scatter diagram for the two sets of ranks. However, extreme caution is necessary in interpreting such a result as implying a causal relationship between the two variables.

Consider the engine size and fuel consumption of Example 9.5. It is known that the greater the engine size of a car, the more power is available. Consequently the fuel consumption increases (corresponding to a decrease in the number of kilometres travelled per litre of fuel). Interpreting this as a causal relationship is reasonable in this instance (that is, increasing the engine size of a car *does* affect the fuel consumption). On the other hand, consider, for example, the average annual salary of teachers and the number of juvenile crimes over a period of several successive years. Both these variables have increased in value over the years and their rank correlation coefficient will have a value fairly close to $+1$, indicating a strong linear association between them. But no one would suggest that the increase in either variable is a direct cause of the increase in the other variable.

A calculated rank correlation coefficient having a value (positive or negative) close to zero implies that there is no *linear* association between the two variables. But note that there may be a curvilinear association, which may be seen from the scatter diagram for the variable values. This emphasises the importance of drawing the scatter diagram before calculating the correlation coefficient. Figure 9.13 shows the scatter diagram for two variables, x and y, for which the correlation coefficient will have a value close to zero, even though a clear curvilinear pattern is displayed by the plotted points. In this case, concluding that there is no association between x and y on the basis of a numerically small correlation coefficient is clearly incorrect.

4 Table 9.19 shows the marks awarded by two judges to six entries in a particular class at a flower show. Convert the marks into ranks and calculate a measure of the extent of agreement between the two judges.

Table 9.19

| Judge A marks | 48 | 50 | 55 | 51 | 47 | 48 |
| Judge B marks | 36 | 38 | 58 | 44 | 52 | 28 |

5 Table 9.20 gives the wing lengths (x cm) and the tail lengths (y cm) of 10 birds of a certain species. Rank the tabulated values and calculate Spearman's rank correlation coefficient. Interpret the result.

Table 9.20

x	y
10.4	7.4
10.8	7.6
10.2	7.2
10.3	7.4
10.2	7.1
10.7	7.4
10.8	7.8
11.2	7.7
10.6	7.8
11.4	8.3

6 Table 9.21 shows the positions at the end of a season in a league competition involving eight football teams, and the average attendances at their home matches during the season. Rank the average home attendances, calculate the rank correlation coefficient and interpret the result.

Table 9.21

Club	A	B	C	D	E	F	G	H
Position	1	2	3	4	5	6	7	8
Average attendance	300	320	120	190	270	180	150	250

7 Table 9.22 shows the latitudes, and the summertime high temperatures and low temperatures of 10 cities in the northern hemisphere. Calculate and interpret the value of the rank correlation coefficient between:
(a) latitude and high temperature,
(b) latitude and low temperature,
(c) high temperature and low temperature.

Table 9.22

City	Latitude (°N)	Temperature (°C) High	Temperature (°C) Low
1	17	32	25
2	6	30	25
3	37	24	22
4	52	12	8
5	45	17	7
6	53	13	4
7	5	26	22
8	19	30	23
9	44	17	6
10	22	32	21

9.7 The Product Moment Correlation Coefficient*

Given observed pairs of values of two quantitative variables, the value of Spearman's rank correlation can only be regarded as a guide to the extent to which the actual observed variable values are linearly associated. For a direct measure of the extent of the linear association between pairs of observed values, Pearson suggested using the **product moment correlation coefficient** r define by:

$$r = \frac{\Sigma(x-\bar{x})(y-\bar{y})}{\sqrt{(\Sigma(x-\bar{x})^2)(\Sigma(y-\bar{y})^2)}}. \qquad (9.5)$$

Here $\bar{x}$ and $\bar{y}$ are the means of the observed x-values and the observed y-values respectively, and the summations (Σ) extend over all the pairs of observed values (x, y). An alternative equivalent form for r which is more convenient when calculating its value is:

$$r = \frac{n\Sigma xy - (\Sigma x)(\Sigma y)}{\sqrt{(n\Sigma x^2 - (\Sigma x)^2)(n\Sigma y^2 - (\Sigma y)^2)}}, \qquad (9.6)$$

where n is the number of pairs of observations. (In an examination, either or both of these expressions for r will probably be given in the information booklet or in the paper itself.)

It may be shown that if the values of the variables are ranked as described in the previous section then the resulting value of r calculated from either of the two expressions will be identical to the value of r_S. It may also be shown that r has precisely the same properties as r_S given in the previous section, namely:

(1) r will always have a value in the interval from -1 to $+1$, inclusive;
(2) $r = +1$ only if the plotted points fall *exactly* on a straight line having positive slope;
(3) $r = -1$ only if the plotted points fall *exactly* on a straight line of negative slope;
(4) $r = 0$ implies that the variables are uncorrelated, in the sense that the plotted points do not show any linear association.

The calculation of r is illustrated in the following examples.

○ **EXAMPLE 9.10**

(Continuation of Examples 9.1 and 9.6)

Calculate the product moment correlation coefficient for the Paper 1 marks and Paper 2 marks given in Table 9.1.

From Table 9.1, the marks on Paper 1, x, and on Paper 2, y, obtained by the 10 pupils are as shown in Table 9.23.

Table 9.23

x	71	67	52	64	44	93	66	85	75	70
y	55	56	47	50	32	66	42	72	54	48

Here $n = 10$, and:

$\Sigma x = 71 + 67 + \ldots + 75 + 70 = 687$,
$\Sigma y = 55 + 56 + \ldots 54 + 48 = 522$,
$\Sigma x^2 = 71^2 + 67^2 + \ldots + 75^2 + 70^2 = 49\,021$,
$\Sigma y^2 = 55^2 + 56^2 + \ldots + 54^2 + 48^2 = 28\,418$,
$\Sigma xy = (71 \times 55) + \ldots + (70 \times 48) = 37\,149$.

Substituting these values in Expression (9.6) we have:

$$r = \frac{10 \times 37\,149 - (687 \times 522)}{\sqrt{(10 \times 49\,021 - 687^2)(10 \times 28\,418 - 522^2)}}$$
$$= \frac{12\,876}{(18\,241 \times 11\,696)} = 0.88 \quad \text{(to 2 d.p.)}.$$

(Recall that the value of the rank correlation coefficient was calculated to be 0.83 in Example 9.6.) Because the calculated value of r is close to $+1$, we conclude that there is a strong positive correlation between the marks on the two papers, a conclusion which is equally apparent from Figure 9.1.

○ **EXAMPLE 9.11**

(Continuation of Example 9.2)

Table 9.24 gives the shoe sizes, x, and the heights, y cm, of 20 boys in a class; these were originally displayed in Table 9.2. Calculate the product moment correlation coefficient.

Table 9.24

x	y	x	y	x	y	x	y
5	159	$5\frac{1}{2}$	165	$6\frac{1}{2}$	168	$7\frac{1}{2}$	172
5	162	6	166	7	172	$7\frac{1}{2}$	173
5	163	6	166	7	170	$7\frac{1}{2}$	174
$5\frac{1}{2}$	167	$6\frac{1}{2}$	170	7	171	8	173
$5\frac{1}{2}$	162	$6\frac{1}{2}$	170	$7\frac{1}{2}$	171	8	175

Here $n = 20$, and we find:

$\Sigma x = 5 + 5 \ldots + 8 + 8 = 130$,
$\Sigma y = 159 + 162 + \ldots + 173 + 175 = 3369$
$\Sigma x^2 = 5^2 + 5^2 + \ldots + 8^2 + 8^2 = 864.5$
$\Sigma y^2 = 159^2 + 162^2 + \ldots + 173^2 + 175^2 = 567\,897$,
$\Sigma xy = (5 \times 159) + (5 \times 162) + \ldots + (8 \times 173) + (8 \times 175) = 21\,981$.

Using Expression (9.6) the product moment coefficient is:

$$r = \frac{20 \times 21\,981 - (130 \times 3369)}{\sqrt{(20 \times 864.5 - 130^2)(20 \times 567\,897 - 3369^2)}}$$
$$= \frac{1650}{\sqrt{(390 \times 7779)}} = 0.95 \quad \text{(to 2 d.p.)}.$$

This is very close to +1, indicating that there is a very strong linear association between the boys' shoe sizes and their heights. (See Figure 9.2.)

○ EXAMPLE 9.12
(Continuation of Examples 9.5 and 9.8)

Table 9.25 shows the engine sizes, x cc, and the fuel consumptions, y km per litre, of 7 different car models; these were originally given in Table 9.5. Calculate the product moment correlation coefficient.

Table 9.25

x	1200	1400	1500	1600	1800	2000	2200
y	14.9	12.7	11.1	89.9	8.1	7.6	7.2

Here $n=7$, and we find:

$\Sigma x = 1200 + 1400 + \ldots + 2000 + 2200 = 11\,700$,
$\Sigma y = 14.9 + 12.7 + \ldots + 7.2 = 71.5$,
$\Sigma x^2 = 1200^2 + \ldots + 2200^2 = 20{,}290{,}000$,
$\Sigma y^2 = 14.9^2 + 12.7^2 + \ldots 7.6^2 + 7.2^2 = 779.73$,
$\Sigma xy = 1200 \times 14.9 + \ldots + 2200 \times 7.2 = 113{,}770$.

Using Expression (9.6) the product moment correlation coefficient is:

$$r = \frac{(7 \times 113\,770) - (11\,700 \times 71.5)}{\sqrt{(7 \times 20\,290\,000 - 11\,700^2)(7 \times 779.73 - 71.5^2)}}$$
$$= -\frac{40\,160}{\sqrt{(5\,140\,000 \times 345.86)}} = -0.95 \quad \text{(to 2 d.p.)}.$$

This is close enough to -1 to indicate a strong linear association between engine size and fuel consumption. But there is some evidence in Figure 9.5 that whereas the association is distinctly linear for engine sizes from 1200 cc to 1800 cc, there is some curvature in the association for engine sizes greater than 1800 cc. Recall that the value of the rank correlation coefficient for the data was -1. (See Example 9.8).

○ EXERCISE 9.6

1 Using the data in Table 9.20, p. 120, calculate the product moment correlation coefficient for the wing lengths and tail lengths of the 10 birds.
2 Using the data in Table 9.22, p. 120, calculate the product moment correlation coefficient for:
(a) latitude and high temperature,
(b) latitude and low temperature,
(c) high temperature and low temperature.
Comment on the values you obtain and on how they compare with the answers you obtained in Question 7, Exercise 9.5.
3 Show that the product moment correlation coefficient for the heights and weights of 20 boys as displayed in Table 9.3, p. 106, is approximately 0.9.
4 Calculate the product moment correlation coefficients for the two sets of values of *x* and *y* given in Questions 5 and 6, Exercise 9.1, p. 108.

Projects

1 Collect various measurements on at least 10 boys (or girls) of roughly the same age. In addition to those we have considered in this chapter, other possible measurements include: outstretched arm length (from armpit to tip of the middle finger), inside leg length, girth of waist, girth of neck, head height (from tip of chin to the top of the head), and so on. Use the methods of this chapter to investigate possible associations between pairs of these measurements.
2 Experiments in Physics and in Biology often involve taking measurements of two (or more) variables. If you are studying either or both of these subjects you should be able to obtain data on two variables. Use the methods of this chapter to investigate whether there is an association between the values you have obtained. (It often happens, especially in Physics, that two variables are known from theory to be linearly related; sometimes the variables have to be transformed for the relation connecting the new variables to be linear. Experiments designed to demonstrate such a theory inevitably produce results which do not agree exactly with the theory because of measurement errors and because the conditions under which the theoretical model holds cannot be controlled adequately.)
3 Published data, such as those contained in H.M.S.O. publications, provide many opportunities for investigating possible associations between pairs of variables. For example, you can obtain data on the number of licensed motor vehicles and the number of casualties (and fatalities) on the roads for each of several successive years by referring to current and previous issues of the annual *Abstract of Statistics*, or the monthly *Digest of Statistics*.

Review Exercises Chapter 9

☐ LEVEL 1

1 Table 9.26 shows duplicate measurements of the volume of a given mass of gas at each of four temperatures.
(a) Draw a scatter diagram of the data.
(b) Calculate the values of $\bar{x}$ and $\bar{y}$, and mark the point $(\bar{x}, \bar{y})$ on your diagram.
(c) Fit a straight line, through $(\bar{x}, \bar{y})$, to the plotted points.
(d) Use your fitted line to estimate (i) the volume of the gas at temperature 65°C, (ii) the increase in the volume of the gas if the temperature is increased by 5°C.
(Keep your answers for use in Question 12.)

Table 9.26

Temperature (x°C)	40	50	60	70
Volume of gas (y cm^2)	1.15	1.18	1.22	1.27
	1.13	1.19	1.24	1.28

2 Table 9.27 gives experimentally observed values of two variables x and y.
(a) Draw a scatter diagram of the pairs of values.
(b) Mark the location of $(\bar{x}, \bar{y})$ on your diagram.
(c) Fit a straight line, through $(\bar{x}, \bar{y})$, to the plotted points.
(d) Use your fitted line to estimate (i) the value of y when x=28, (ii) the decrease in the value of y when x is increased by 10 units.
(Keep your answers for use in Question 13.)

Table 9.27

x	5	10	15	20	25	30	35	40
y	9.4	8.8	8.1	7.4	6.9	6.1	5.3	4.8

3 Table 9.28 shows the index numbers for production and prices for a particular commodity over 10 successive years.
(a) Draw the scatter diagram for the production index and price index.
(b) Calculate $(\bar{x}, \bar{y})$ and plot it on your diagram.
(c) Fit a straight line to your plotted points. Use it to estimate the price index when the production index is 100 and the production index when the price index is 120. State any reservations you may have about the reliability of each of your estimates.
(d) Calculate and interpret the value of the rank correlation coefficient for the production index and price index.
(Keep your answers for use in Question 14).

Table 9.28

Year	Production index (x)	Price index (y)
1	111	94
2	113	98
3	96	105
4	95	110
5	91	111
6	86	110
7	102	98
8	108	105
9	89	111
10	99	105

4 Table 9.29 shows the positions of eight soccer teams at the end of the league programmes in 1984 and 1985. Calculate the rank correlation coefficient and comment on its value.

Table 9.29

Team	A	B	C	D	E	F	G	H
1984 position	1	2	3	4	5	6	7	8
1985 position	3	5	4	2	7	1	6	8

5 Two motoring correspondents, A and B, ranked nine new car models as shown in Table 9.30. Calculate a measure of the extent of the agreement between the two correspondents and comment on its value.

Table 9.30

Model	A's ranking	B's ranking
I	2	1
II	3	2
III	6	4
IV	1	3
V	5	7
VI	4	6
VII	9	5
VIII	7	8
IX	8	9

6 Two estate agents were asked to rank seven houses according to their market values, from highest to lowest. The ranking of the actual market values and those assigned by the two estate agents are given in Table 9.31. Determine which of the two estate agents gave ranks closest to the those of the actual ranks.

[123

Table 9.31

Actual ranking	1	2	3	4	5	6	7
A's ranking	2	1	4	5	3	7	6
B's ranking	3	2	1	4	8	5	6

7 Table 9.32 shows the amount of coal consumed (x thousand tonnes) and the production of pig iron (y thousand tonnes) at each of six factories.
(a) Draw a scatter diagram of coal consumption and pig iron production, and on it mark the location of $(\bar{x}, \bar{y})$.
(b) Fit a straight line to the plotted points. Use your fitted line to estimate a representative value for pig production when the coal consumption is 9300 tonnes, and a representative value for coal consumption when the pig iron production is 5000 tonnes.
(c) Calculate the rank correlation coefficient. What feature of the scatter diagram is being measured approximately by the rank correlation coefficient? (Keep your answer for use in Question 16.)

Table 9.32

Factory	1	2	3	4	5	6
Coal consumption (x)	14	12	7	6	8	10
Pig iron production (y)	7	6	4	3	4	6

8 Sketch diagrams of ordinal (rank) variables x and y for which the value of the rank correlation coefficient r_S is (a) approximately 0, (b) exactly +1, (c) exactly −1, (d) a positive number between $\frac{1}{2}$ and 1.

9 The number of vehicles licensed in millions, and the number of road accidents, in thousands, in a certain country over a period of seven years are displayed in Table 9.33.

Table 9.33

Year	No. of vehicles (m)	No. of accidents ('000)
1	2·0	73
2	2.3	77
3	2.4	78
4	2.5	80
5	2.8	83
6	3.0	86
7	3.2	90

(a) Draw a scatter diagram, with the number of vehicles along the horizontal axis and the number of accidents along the vertical axis. State the scales you have used. Plot the point corresponding to the mean values of the two variables. Fit a straight line to the plotted points.
(b) Calculate the slope of the line you fitted in (a).
(c) Use your diagram to forecast the number of accidents in the next year given that the number of licensed vehicles increases to 3.3 million.

10 The ages and the selling prices of five used cars of a particular model are given in Table 9.34.
(a) Draw a scatter diagram of selling price, y, against age x. Plot the point $(\bar{x}, \bar{y})$ on your diagram. (b) Draw a straight line through $(\bar{x}, \bar{y})$ to fit the plotted points in your diagram.
(c) Determine the equation of your fitted line and use it to estimate a representative selling price for cars of the same model that are 6 years old.

Table 9.34

Age (years)	1	2	3	4	8
Price (£'000)	6.5	5.8	5.2	4.4	1.9

11 The heights and the yields of five tomato plants were measured, with the results shown in Table 9.35.
(a) Draw a scatter diagram for the data.
(b) Fit a straight line to the plotted points and determine its equation.
(c) Use your equation to estimate a representative yield from plants of height 1.25 m and a representative height for plants yielding 1.30 kg.

Table 9.35

Height (m)	0.95	0.60	1.35	0.65	0.90
Yield (kg)	1.90	1.10	2.95	1.65	2.50

☐ **LEVEL 2**

12 For the data in Table 9.26, p. 123 find the equation of the least squares regression line of volume of gas on temperature. Use it to estimate the mean volume of the gas when the temperature is 65°C, and compare your answer with that you obtained in Question 1(d)(i).

13 For the data in Table 9.27, p. 123:
(a) find the equation of the 'best' line for estimating the mean value of y for a specified value of x. Use it to estimate the mean value of y when x=28,

comparing it with the estimate you obtained in Question 2(d)(i);

(b) calculate and interpret the value of the product moment correlation for x and y.

14 Calculate and interpret the value of the product moment correlation coefficient for the production and price indexes given in Table 9.28. Comment on its value relative to the value of the rank correlation coefficient you calculated in Question 3(d).

15 Draw a scatter diagram of the heights, x, and the weights, y, of 16 girls as given in the table you constructed from Table 1.2 in Question 2, Exercise 9.1. Find the equation of the 'best' line for estimating the mean weight of girls of a given height; hence estimate the mean weight of girls of height 150 cm.

16 Calculate and interpret the value of the product moment correlation coefficient of the coal consumptions and the pig iron productions given in Table 9.32, p. 124. Comment on its value relative to that of the rank correlation coefficient you calculated in Question 7(c).

17 The lengths, to the nearest cm, and the weights, to the nearest 5 g, of six fish of a certain species are given in Table 9.36

Calculate and interpret the value of the product moment correlation coefficient for the lengths and the weights of the six fish.

Table 9.36

Length (x cm)	28	22	32	25	35	29
Weight (y g)	490	480	495	485	500	490

10

TIME SERIES

"When a man sits with a
pretty girl for an hour, it seems like
a minute. But let him sit on
a hot stove for a minute and it's
longer than any hour."

Einstein, quoted in his obituary,
The New York Times (*April 19, 1955*)

10.1 Time Series Data

Time series data are observations of a variable over a period of time. In most cases the observations are made at regular time intervals, which may be hourly, daily, weekly, monthly, quarterly, or annually. Weather stations keep daily records of temperatures, sunshine hours, and rainfall. Various government departments collect and publish data for months, quarters, and years on variables such as the rate of inflation, number unemployed, population size, energy consumption, and many more. Retailers and manufacturers keep records of their weekly, monthly and annual sales and profits.

○ **EXAMPLE 10.1**

Table 10.1, extracted from Issue No. 484 of the monthly *Digest of Statistics* shows U.K. consumers' expenditures on tobacco, in millions of pounds, during each of the years from 1980 to 1985.

The data are displayed diagrammatically in Figure 10.1, which is called a **line chart** or a **line graph**. In this diagram the horizontal axis represents time and the vertical axis represents the variable. The data

Table 10.1 *U.K. Consumers' Annual Expenditure on Tobacco*

Year	1980	1981	1982	1983	1984	1985
Expenditure (£m)	4822	5515	5882	6208	6622	7010

points have been plotted and joined up with broken straight lines to provide a pictorial display of the changes in the variable (annual expenditure) over the years. Broken lines are appropriate because points between those plotted have no meaningful interpretation. For example, annual expenditure at some value along the horizontal axis in between two successive years does not exist.

It is clear from Figure 10.1 that the annual expenditure on tobacco has increased throughout the period. Since the data are expenditures it does not necessarily follow that the amount of tobacco sold

Figure 10.1 *Line Chart for the Data in Table 10.1*

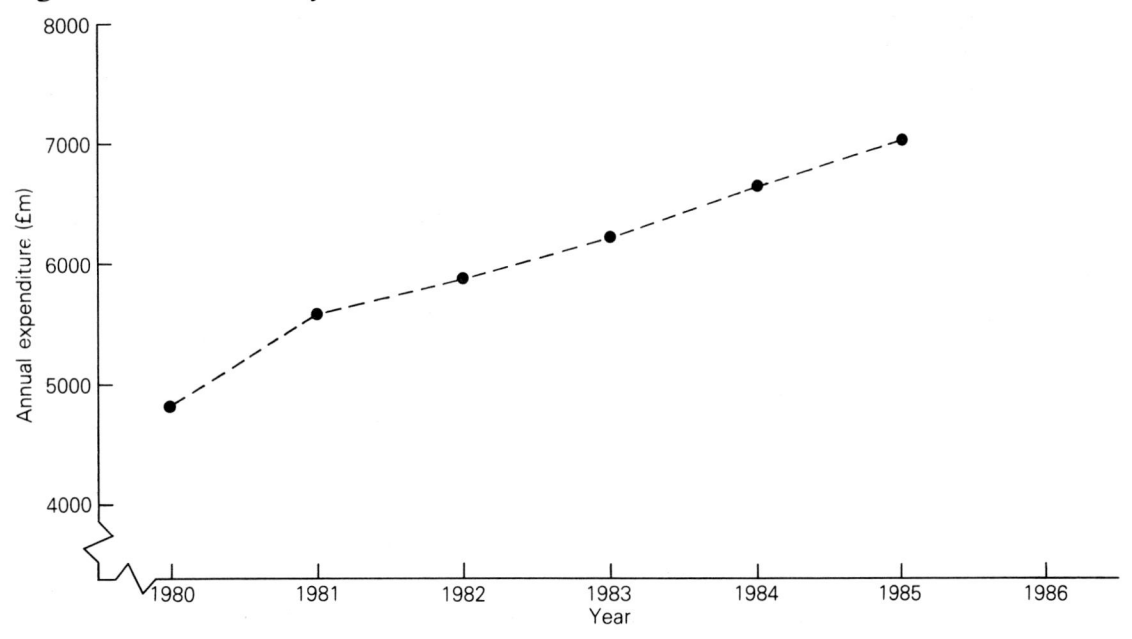

Figure 10.2 *Line Fitted to the Data in Table 10.1*

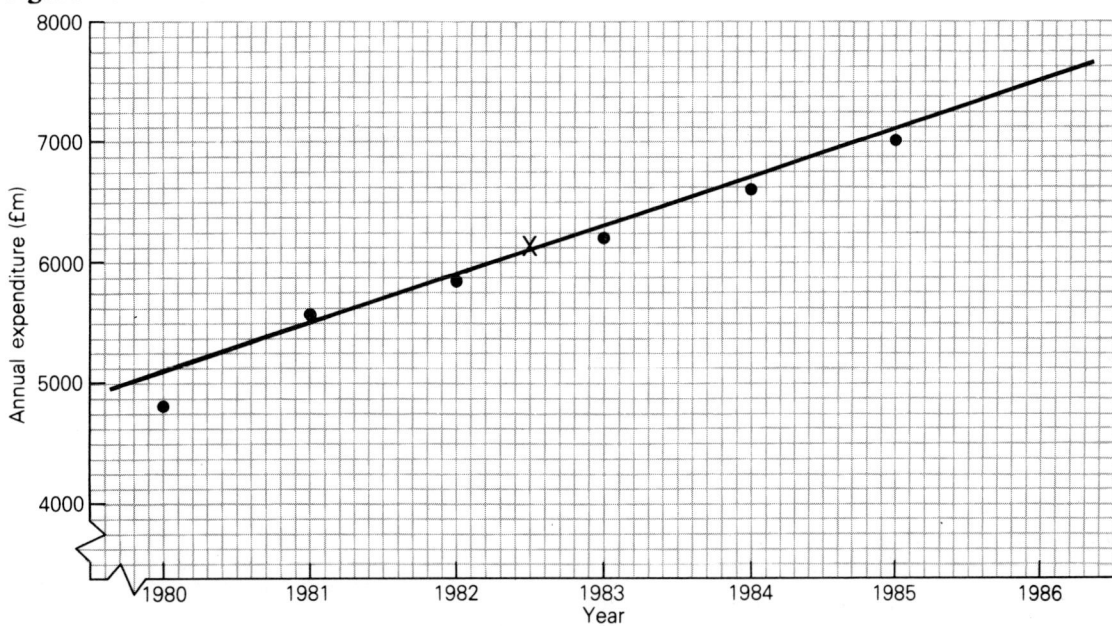

per year has increased over the period. This is because no allowance has been made for the varying cost of tobacco over the period. From the same issue of the *Monthly Digest of Statistics* we find that the yearly expenditures, in millions of pounds *revalued at 1980 prices*, for the years from 1980 to 1985 are, respectively:

4822, 4470, 4128, 4083, 3944, 3836.

These figures show that the amount of tobacco sold per year has actually decreased over the period.

One purpose for collecting time series data is to make **predictions** or **forecasts**. Using the data in our example let us predict the 1986 expenditure on tobacco in the U.K. To do so we fit a line to the plotted points. This is called the **trend line**. The points shown in Figure 10.1 suggest that a straight line fit is reasonable. We indicated in Chapter 9 that to fit a straight line there are advantages in making the fitted line pass through the mean of the data. In the present example the mean time is simply the mid-point of the interval from 1980 to 1985, which is the time point midway between 1982 and 1983. The mean annual expenditure is:

$$\frac{4822+5515+5882+6208+7010}{6} = 6010,$$

[127

to the nearest million pounds. In Figure 10.2 the mean is indicated by a cross (**X**), and the straight line shown is one fitted 'by eye' to the observed points so as to pass through the mean. The line has been extended to 1986 to give us an estimate (prediction) of the tobacco expenditure in 1986.

Reading from the graph, our estimate of the 1986 expenditure is about £7500m. This prediction is based on the assumption that the trend during 1980–1985 will continue into 1986. This is reasonable provided no unforeseen circumstances arise to change the trend. This might happen, for example, if the government decided to ban the sale of tobacco or to introduce a very much higher tax on tobacco.

○ **EXAMPLE 10.2**

Table 10.2, extracted from the 1986 Edition of the *Annual Abstract of Statistics*, shows the numbers of notified cases of measles in England and Wales, in thousands, in each of the years from 1978 to 1984.

The line graph for the data is displayed in Figure 10.3. Unlike Example 10.1, no apparent trend is displayed by the data over the period, so that it is not worthwhile to make any attempt at prediction.

Table 10.2 *Yearly Numbers of Notified Cases of Measles in England and Wales*

Year	Cases ('000)
1978	124.1
1979	77.4
1980	139.5
1981	53.0
1982	94.2
1983	103.7
1984	62.1

Figure 10.3 *Line Chart for the Data in Table 10.2*

● **EXERCISE 10.1**

1 *Table 10.3, extracted from the 1986 Edition of* Social Trends, *shows the number of driving tests (for cars and two wheelers) conducted in each of the years from 1979 to 1984, together with the number that passed in each year.*

(a) Draw a time series graph of the number of tests conducted per year and comment on the trend shown.

(b) Calculate the percentage pass rate for each year. Draw a time series graph of the percentage pass rates and comment.

Table 10.3

Year	No. of tests	No. of passes
1979	1570	742
1980	1962	928
1981	2031	967
1982	2005	965
1983	1892	921
1984	1784	875

2 *Table 10.4, extracted from Issue No. 387,* Economic Trends, *shows U.K. quarterly consumptions of coal and petroleum, in million tonnes of coal or their equivalent, during the years 1983, 1984, and 1985.*

(a) Using the same axes and scales draw the time series graphs of the quarterly consumptions of coal and petroleum.

(b) During the period concerned there had been a coalminers' strike. Use your graph to determine the period of the strike, and comment on the consumption of petroleum during that period.

Table 10.4 *U.K. Quarterly Consumptions of Coal and Petroleum (million tonnes)*

Year and quarter		Coal	Petrol
1983	1	113	103
	2	109	108
	3	116	113
	4	111	102
1984	1	113	103
	2	72	136
	3	63	153
	4	63	150
1985	1	72	143
	2	119	103
	3	113	108

3 Table 10.5, extracted from Issue No. 16 of Social Trends, *shows the life expectation, in years, of a newly born girl.*
(a) Graph the data.
(b) Comment on the trend shown in your graph. Suggest an explanation for this trend.
(c) Plot the mean of the data on your graph.
(d) Fit a straight line trend to the data and use it to obtain an estimate of the life expectation of a girl born in 1991.

Table 10.5

Year of birth	Life expectation (years)
1951	71.2
1961	73.8
1971	75.0
1981	76.2

10.2 Seasonal Variation

Any general pattern that is displayed by the line graph of a particular time series is referred to as the **secular** or **long-term trend**. In Example 10.1 the annual expenditures on tobacco showed an increasing trend which was approximately linear. On the other hand, in Example 10.2 there was no apparent secular trend in the yearly number of notified cases of measles.

Even when there is a general trend the individual variable values may well show considerable variation about that trend. Such variation may arise from several possible sources, some of which we may be able to identify. The variation that cannot be accounted for from the identified sources is usually referred to as **random** or **irregular variation.**

The only identifiable source that we will consider here is that due to **seasonal variation.** For example, in any year ice cream sales will generally be higher in the summer than in the winter, while domestic fuel consumption will generally be higher in the winter than in the summer. Seasonal variation may also occur at times of year which do not coincide with seasons as we know them. For this reason seasonal variation is sometimes referred to as **cyclical variation.** The following example is one in which the data show seasonal (cyclical) variation.

○ **EXAMPLE 10.3**

Table 10.6, extracted from Issue No. 484 of the *Monthly Digest of Statistics*, shows U.K. consumers' expenditures on tobacco, in millions of pounds, in each quarter for the years from 1982 to 1985. (Table 10.1 gave the corresponding annual expenditures.)

Table 10.6 *Quarterly Expenditures on Tobacco in the U.K.*

Year and quarter		Expenditure (£m)	Year and quarter		Expenditure (£m)
1982	1	1396	1984	1	1568
	2	1460		2	1640
	3	1496		3	1691
	4	1530		4	1723
1983	1	1475	1985	1	1683
	2	1558		2	1742
	3	1569		3	1776
	4	1606		4	1809

The line chart for the data is shown in Figure 10.4. The general upward trend in expenditure noted in Figure 10.1 is, of course, still evident. But we also see that the pattern within each year is very similar, in that the expenditure increases as we go through the quarters. This cyclical pattern is evidence of the presence of seasonal variation, the 'seasons' being the four quarters of a year. The drops in the expenditure in the first quarters of 1983, 1984, and 1985, as compared with those in the last quarters of the preceding years, are likely to be due to increased expenditure on tobacco over the Christmas periods.

Since the fluctuations in the plotted points are not very large it is not too difficult to fit a straight line to the points by eye to provide an estimate of the general trend. A more appropriate method for

Figure 10.4 *Line Chart for the Data in Table 10.6*

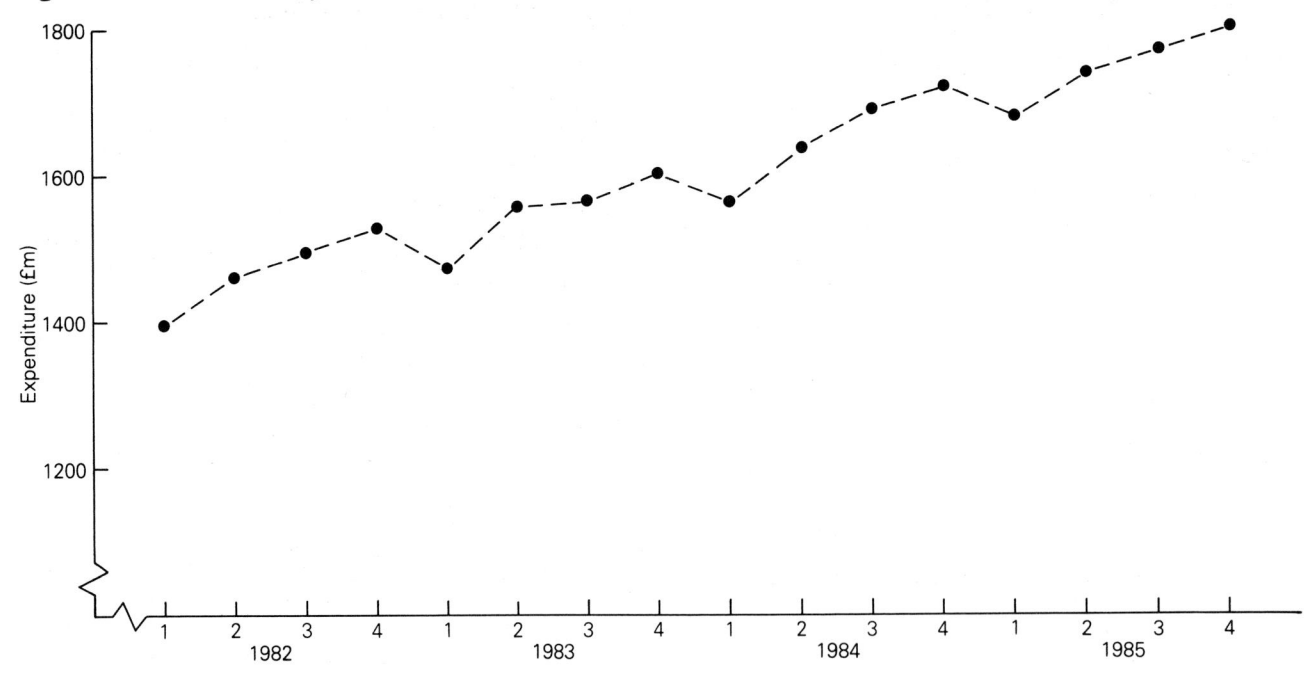

estimating a trend when there is seasonal variation is described in the next section.

● **EXERCISE 10.2**

Display the time series data in each of the following tables as a line chart. Describe any seasonal variation that is present.
Keep your charts for use in Exercise 10.3.

1 *Numbers of Live Births in England and Wales*

Year and quarter		Births ('000)
1982	1	153
	2	157
	3	162
	4	153
1983	1	152
	2	161
	3	163
	4	152
1984	1	154
	2	158
	3	167
	4	158
1985	1	160
	2	165
	3	172
	4	160

2 *U.K. Households' Quarterly Expenditures on Food*

Year and quarter		Expenditure (£m)	Year and quarter		Expenditure (£m)
1982	1	6106	1984	1	6846
	2	6519		2	7235
	3	6335		3	7031
	4	6630		4	7336
1983	1	6394	1985	1	7251
	2	6706		2	7577
	3	6622		3	7293
	4	6900		4	7600

3 *Quarterly Sales by a Manufacturing Company*

Year and quarter		Sales (£'000)
1983	1	27
	2	40
	3	31
	4	42
1984	1	52
	2	77
	3	69
	4	89
1985	1	92
	2	123
	3	108
	4	132

10.3 Moving Averages*

When dealing with time series data in which there is seasonal (cyclical) variation, the plotted points on the line chart may show large fluctuations, with upward and downward peaks. Such fluctuations make it more difficult to fit a reasonable trend line over the period. As it happens, the line chart in Figure 10.4 is not too bad in this respect, and it is not difficult to fit a straight line to it by eye to estimate the general trend. But this will not always be the case. To estimate the trend it would be nice if we could generate a new set of points representative of the plotted observed points but having a smoother profile. This can be done by using **moving averages**.

Moving averages are the averages of successive observations in a time series. The averaging may be done over successive pairs, or triples, or quadruples, and so on. When averaging over pairs the first moving average is the average of the first and the second observations in the time series; the second moving average is the average of the second and third observations; the fourth moving average is the average of the third and fourth observations; and so on. The resulting values are referred to as the **two-point moving averages**. These values are then plotted on a graph against the midpoints of the time periods covered by the pairs of observations that were averaged. The plotted moving averages, having smaller fluctuations, give a clearer picture of the general trend than the original data.

Consider the data in Table 10.7

Table 10.7

Time point	1	2	3	4	5	6	7	8
Observation	20	28	32	26	30	34	38	29

For this time series the two-point moving averages and their associated time points are:

    1st moving average: $\frac{1}{2}(20+28)=24$
       time point: $\frac{1}{2}(1+2)=1\frac{1}{2}$,
    2nd moving average: $\frac{1}{2}(28+32)=30$,
       time point: $\frac{1}{2}(2+3)=2\frac{1}{2}$,
    3rd moving average: $\frac{1}{2}(32+26)=29$,
       time point: $\frac{1}{2}(3+4)=3\frac{1}{2}$,

and so on, the final one being:

    7th moving average: $\frac{1}{2}(38+29)=33\frac{1}{2}$,
       time point: $\frac{1}{2}(7+8)=7\frac{1}{2}$.

Similarly, taking three successive observations in turn the **three-point moving averages** and associated time points for the data in Table 10.7 are:

    1st moving average: $(20+28+32)/3=26.67$,
       time point: $(1+2+3)/3=2$,
    2nd moving average: $(28+32+26)/3=28.67$,
       time point: $(2+3+4)/3=3$,

and so on, the final one being:

    6th moving average: $(34+38+29)/3=33.67$,
       time points: $(6+7+8)/3=7$.

When there is seasonal (cyclical) variation present in a time series it may be shown that the most effective moving averages for estimating the general trend are those taken over as many observations as there are in one complete cycle. For example, in Figure 10.4 a complete cycle is one year, since the pattern of the line chart is similar in each year. In this case four-point moving averages should be used to estimate the general trend. Since the time points are quarters these four-point moving averages are, in fact, four-quarter moving averages.

○ **EXAMPLE 10.3** *(Continued)*

Determine the four-quarter moving averages for the data in Table 10.6 and plot them on a graph. Fit a straight line to the plotted values by eye to estimate the general trend of the data.

The calculation of the four-point moving averages is detailed in the Table 10.8.

Each entry in the third column of Table 10.8 is the sum of four observed expenditure values, working upwards from the observed expenditure value in the same row as the entered total. A convenient way of obtaining these totals when preparing such a table is as follows. Place a ruler (or a sheet of paper) with its upper edge just below the row containing the fourth observed expenditure value. This is 1530 in our table. Now total the four observed expenditure values above the ruler and enter the result, 5882, in the third column. Now move the ruler's upper edge down one row so as to be immediately below the fifth observed expenditure value, 1475. Place the lower edge of another ruler immediately above the second observed expenditure value, 1460. You then have four values, from the second to the fifth, between the two rulers. Sum these four values and enter the result, 5961, in the third column. Moving both rulers down one row and summing the four observed expenditure values between them gives the third entry, 6059, in the third column. Continue in this way until all the entries in the third column have been obtained.

The four-quarter moving averages in the fourth column of the table are obtained by dividing the

[131

Table 10.8

Year and quarter	Observed expenditure (£m)	Four-quarter total (£m)	Four-quarter moving average (£m)	Associated time point
1982 1	1396			
2	1460			
			$1470\frac{1}{2}$	1982: $2\frac{1}{2}$
3	1496			
			$1490\frac{1}{4}$	1982: $3\frac{1}{2}$
4	1530	5882		
			$1514\frac{3}{4}$	1983: $\frac{1}{2}$
1983 1	1475	5961		
			1533	1983: $1\frac{1}{2}$
2	1558	6059		
			1552	1983: $2\frac{1}{2}$
3	1569	6132		
			$1575\frac{1}{4}$	1983: $3\frac{1}{2}$
4	1606	6208		
			$1595\frac{3}{4}$	1984: $\frac{1}{2}$
1984 1	1568	6301		
			$1626\frac{1}{4}$	1984: $1\frac{1}{2}$
2	1640	6383		
			$1655\frac{1}{2}$	1984: $2\frac{1}{2}$
3	1691	6505		
			$1684\frac{1}{4}$	1984: $3\frac{1}{2}$
4	1723	6622		
			$1709\frac{3}{4}$	1985: $\frac{1}{2}$
1985 1	1683	6737		
			1731	1985: $1\frac{1}{2}$
2	1742	6839		
			$1752\frac{1}{2}$	1985: $2\frac{1}{2}$
3	1776	6924		
4	1809	7010		

Figure 10.5 *Four-Quarter Moving Averages for the Data in Table 10.6*

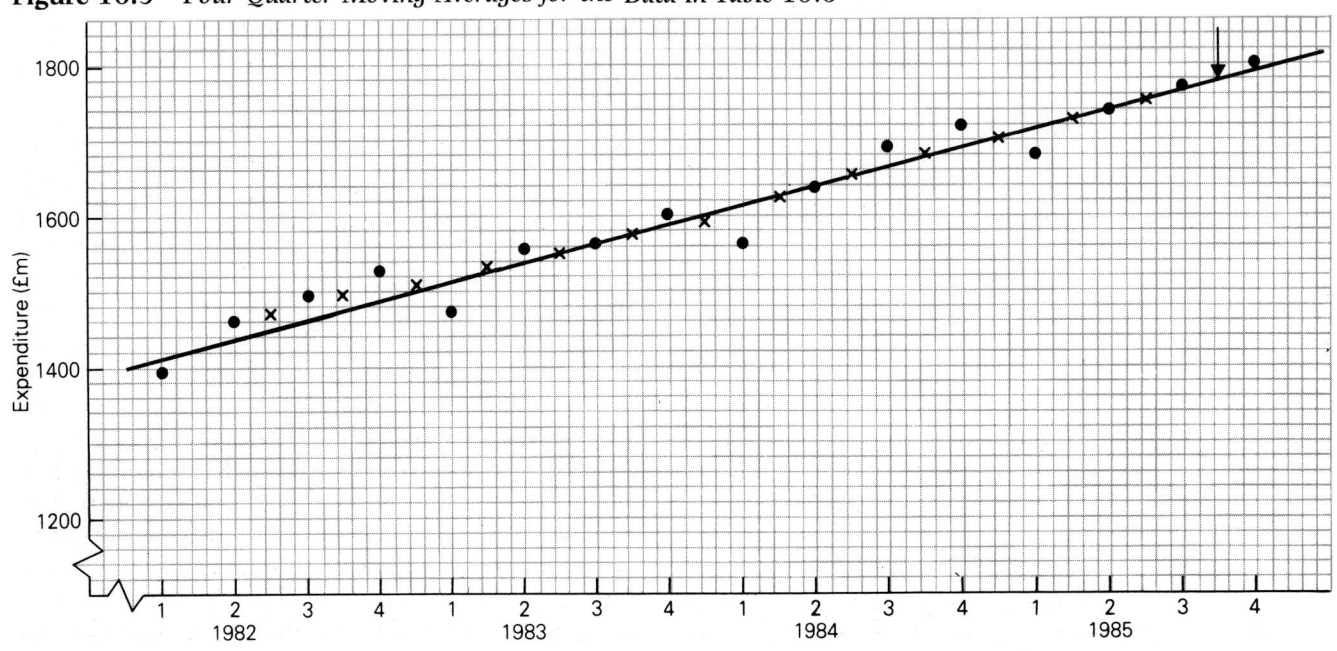

ELECTION FEVER

In the run-up to the 1987 general election in the United Kingdom, opinion polls and "polls of polls" preoccupied the media. But what is the reaction to such polls, and how are they interpreted? Should the results of different polls, which depended on different questions asked in different ways to different people, be combined? Does it make a difference if the responses are collected by phone, in person, or by questionnaire?

Early Election Indicators
(Averages of Interview Polls)

	After first 5 days	After second 5 days	
Conservative			50% / 40%
Labour			30%
Alliance			20%

Key
―――― 1983 – 3 polls   ―·―·― 1983 – 6 polls
-------- 1987 – 7 polls   ·········· 1987 – 3 polls

The data are like time series data, and the media were full of "trend interpretations" based on attempts to smooth out the random fluctuations between successive polls. Meanwhile, the Party leaders saw nothing but encouragement. They talked about the fluctuations as if they were trends, and trends as if they were fluctuations, whichever made their cases sound better.

Infact, the Alliance and Labour parties did worse than the early "trends" would have predicted. The Conservatives did better. There were, of course, still so-called "experts" who could find different "interpretations" of the data. With more than 200 polls in the three weeks immediately before the election, how is the ordinary voter supposed to cope with such overdoses of data? How influential do you think the collection and presentation of such data are on the final outcome?

entries in the third column by four. These values have been positioned in the table to correspond with the midpoints of the time intervals over which the averaging has been done. The midpoints are shown in the last column of the table.

In Figure 10.5 we have plotted the original observations, shown by dots (●), and the moving averages 'from Table 10.8, shown by crosses (**X**). It is clear that the moving averages have a smoother profile than the original observations. Joining up the plotted moving averages provides an estimate of the general trend. Here the trend is approximately linear; the straight line shown in Figure 10.5 is one fitted by eye to the plotted moving averages. (In this example we have chosen not to use the technique described in Chapter 9 of fitting the line so as to pass through the mean.)

□□ PREDICTION

The fitted trend line to the moving averages can be used to predict (estimate) the value of the variable that will be observed at some future time point. In the previous example suppose we want to predict the next observation, namely tobacco expenditure in the first quarter of 1986. To do so we first use the fitted line to estimate the value of the next moving average. This will be the value on the fitted line corresponding to the time point 1985: $3\frac{1}{2}$, which is the midpoint of the time period over which the average is calculated. From Figure 10.5 the required estimate is about 1782 (arrowed in the figure).

Now this particular moving average would be calculated by averaging the expenditures from the second quarter in 1985 to the first quarter in 1986, inclusive. Denoting the estimated expenditure in the first quarter of 1986 by e and using the observed expenditures for the last three quarters in 1985, the value of e must be such that:

$$\tfrac{1}{4}(1742+1776+1809+e) \approx 1782,$$

from which we find that:

$$e \approx (4 \times 1782) - (1742+1776+1809) = 1801.$$

Our estimate of the tobacco expenditure in the first quarter of 1986 is £1801 million.

○ EXAMPLE 10.4

Table 10.9 shows the sales, in pence, of ice cream from a van parked on a promenade on each day of three successive weeks during a summer. Calculate and plot on a graph moving averages appropriate for estimating the trend of the sales over the period. Estimate the sales of ice cream on the Sunday of the fourth week.

Table 10.9 *Daily Sales of Ice Cream (pence)*

Day	Sun	Mon	Tue	Wed	Thu	Fri	Sat
Week 1	950	510	530	580	500	480	870
Week 2	910	490	490	600	480	450	840
Week 3	860	430	490	580	460	420	790

The line chart for the data in Table 10.9 is displayed in Figure 10.6. Here we see that there are large fluctuations in successive observations. However, the pattern of the sales is very similar in each of the three weeks. This means that there is seasonal or cyclical variation, a complete cycle being one week. We should therefore operate with seven-point moving averages in order to obtain an estimate of the general trend.

The calculation of the seven-point moving averages, to the nearest penny, is shown in Table 10.10. In Figure 10.7 we have plotted the original observations (dots) and the moving averages (crosses), and fitted a straight line by eye to the plotted moving averages as an estimate of the general trend.

Figure 10.6 *Line Chart for the Data in Table 10.9*

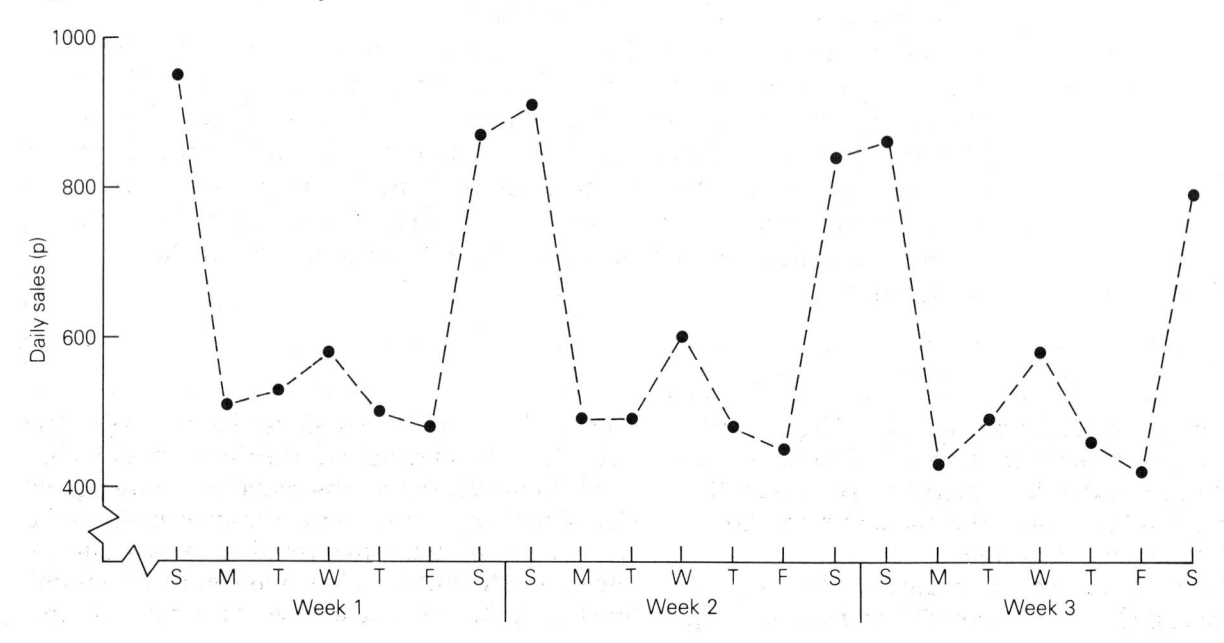

Figure 10.7 *Seven-Point Moving Averages for the Data in Table 10.9*

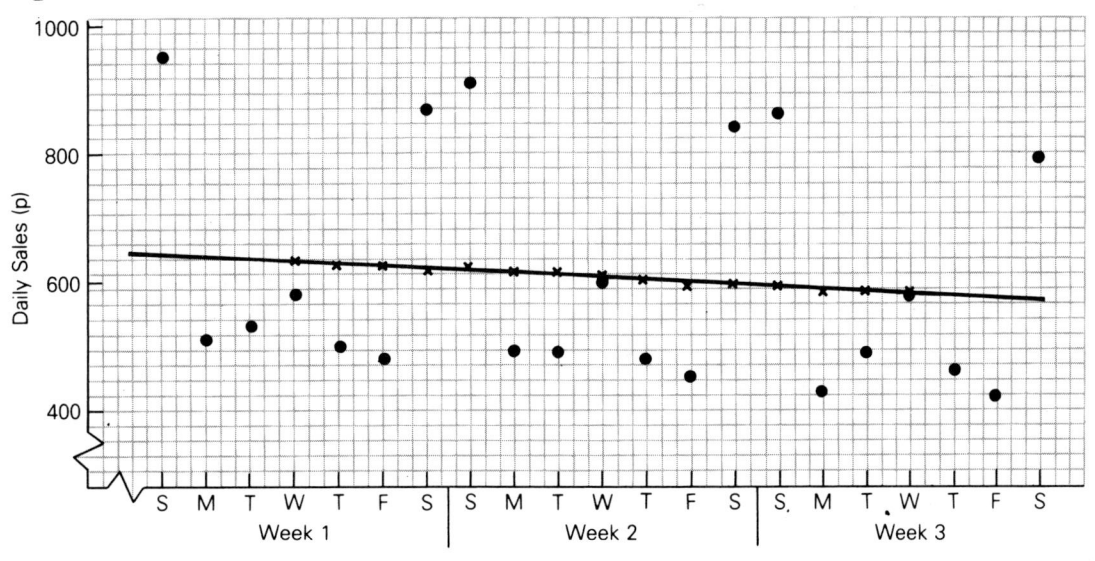

Table 10.10

Week and day	Sales (p)	Seven-day total (p)	Seven-day moving average (p)	Associated time point
1 Sun.	950			
Mon.	510			
Tue.	530			
Wed.	580		631	Wk 1: Wed.
Thu.	500		626	Thu.
Fri.	480		623	Fri.
Sat.	870	4420	617	Sat.
2 Sun.	910	4380	620	Wk 2: Sun.
Mon.	490	4360	617	Mon.
Tue.	490	4320	613	Tue.
Wed.	600	4340	609	Wed.
Thu.	480	4320	601	Thu.
Fri.	450	4290	593	Fri.
Sat.	840	4260	593	Sat.
3 Sun.	860	4210	590	Wk 3: Sun.
Mon.	430	4150	587	Mon.
Tue.	490	4150	583	Tue.
Wed.	580	4130	576	Wed.
Thu.	460	4110		
Fri.	420	4080		
Sat.	790	4030		

Using the fitted line we see that the next moving average, corresponding to the time point Week 3: Thursday, is estimated as 580. Denoting the estimated sales on the next Sunday by e, we have:

$$\tfrac{1}{7}(430+490+580+460+420+790+e) \approx 580,$$

and:

$$e \approx (7 \times 580) - (430+490+580+460+420+790) = 890.$$

If follows that our estimate of the sales of ice cream on the Sunday of the fourth week is 890 pence, or £8.90.

● **EXERCISE 10.3**

(Keep your graphs for use in later exercises.)
Questions 1 and 2. Determine the four-quarter moving averages for the data given in each of Questions 1 and 2, Exercise 10.2, p. 130 and plot them on a graph. Fit a trend line to your plotted points and comment. In each case obtain an estimate of the variable value in the first quarter of 1986.

3 Table 10.11 shows the value of a certain price index in the months of April, August, and December for the years 1982 to 1985.

Table 10.11

	Month		
Year	April	August	December
1982	88	118	124
1983	119	150	154
1984	145	178	183
1985	177	209	215

(a) Plot the data on graph paper.
(b) By plotting appropriate moving averages on your graph obtain a straight line estimate of the general trend over the period.
(c) Find an estimate of the value of the price index in April, 1986.

ESTIMATING SEASONAL EFFECTS

At any given time point the difference:

(Observed value − Trend value)

is an estimate of the seasonal effect at that time point. With reference to Example 10.3, let us find an estimate of the seasonal effect on tobacco expenditure in (1) the first quarter of 1982, (2) the fourth quarter of 1982.

(1) From Table 10.6 the observed expenditure in the first quarter of 1982 was £1396m. From the fitted trend line in Figure 10.5 an estimate of the trend value in the first quarter of 1982 is about £1423m. It follows that the estimated seasonal effect in the first quarter of 1982 is

$$(1396 - 1423) = -27,$$

that is, a reduction of about £27m.

(2) Similarly, from Table 10.6 and Figure 10.5, the estimated seasonal effect in the fourth quarter of 1982 is:

$$(1530 - 1500) = 30,$$

that is, an increase of £30m.

Now consider estimating the *average* seasonal effect on the variable value in a first quarter. This may be estimated by taking the average of the seasonal effects for the first quarters during the period of observation. We have already shown that the estimated seasonal effect in the first quarter of 1982 is −£27m. Reference to Table 10.6 and Figure 10.5 will give the observed and the trend values corresponding to the first quarters of 1983, 1984 and 1985. The results obtained are shown in Table 10.12.

Table 10.12

First quarter of:	1982	1983	1984	1985
Observed value (O)	1396	1475	1568	1683
Trend value (T)	1423	1523	1621	1720
Seasonal effect (O−T)	−27	−48	−53	−37

Thus an estimate of the average seasonal effect on tobacco expenditure during a first quarter is:

$$-\tfrac{1}{4}(27+48+53+37) = -41\tfrac{1}{4},$$

that is, a reduction of about £41m, on average.

4 With reference to Example 10.3 obtain an estimate of the average seasonal effect on tobacco expenditure in a fourth quarter.

5 With reference to Table 10.10 obtain estimates of:
(a) the seasonal effect on ice cream sales on the Saturday of the first week,
(b) the average seasonal effect on Saturday ice cream sales.

6 With reference to Question 1 of this Exercise obtain estimates of:
(a) the seasonal effect on households' expenditures on food in the second quarter of 1982,
(b) the average seasonal effect on households' expenditure on food in a second quarter.

7 With reference to Question 2 of this Exercise obtain estimates of:
(a) the seasonal effect on sales in the fourth quarter of 1985,
(b) the average seasonal effect on sales in a fourth quarter.

8 With reference to Table 10.11 obtain estimates of
(a) the seasonal effect on the value of the price index in August 1982,
(b) the average seasonal effect on the value of the price index in August.

Review Exercises Chapter 10

☐ LEVEL 1

1 Table 10.13, extracted from the 1986 Edition of Social Trends, gives the expectations of life, in years, of boys born in certain years.
(a) Graph, the data. Comment on the trend shown in your graph. Suggest an explanation for this trend.
(b) Plot the mean of the data on your graph.
(c) Draw a straight line estimate of the general trend and use it to predict the life expectation of a boy born in 1991.
(d) Contrast the data in Table 10.13 with the data on girls given in Table 10.5, p. 129.

Table 10.13

Year of birth	1951	1961	1971	1981
Expectation (years)	66.2	67.9	68.8	69.8

2 The numbers, in millions, of all motor vehicles that were licensed in the U.K. during the years from 1978 to 1984 are shown in Table 10.14

extracted from Issue No. 484 of the Monthly Digest of Statistics.
(a) Graph the data.
(b) Plot the mean of the data on your graph.
(c) Fit a straight line passing through the mean to your plotted points. Use your fitted line to estimate the number of licensed motor vehicles in the U.K. in 1985.

Table 10.14 *Numbers of Licensed Motor Vehicles in the U.K. 1978–1984*

Year	1978	1979	1980	1981	1982	1983	1984
No. (m)	17.8	18.6	19.2	19.4	19.8	20.2	20.8

3 Table 10.15, extracted from the 1986 Edition of the Annual Abstract of Statistics, shows the number, in thousands of notified cases of tuberulosis in the U.K. in each of the years from 1979 to 1984.
(a) Draw on graph paper the line chart for the data. Offer a possible explanation for the trend shown.
(b) Plot the mean of the data on your graph. By eye fit a straight line passing through the mean to the plotted points, and use it to estimate the number of cases of tuberculosis that were notified in 1985.

Table 10.15

Year	1979	1980	1981	1982	1983	1984
Number ('000)	10.7	10.5	9.3	8.4	7.8	7.0

4 Table 10.16, extracted from the 1986 Edition of the Annual Abstract of Statistics, shows the average admission price to cinemas in Great Britain in certain years.
(a) Graph the data as a line chart.
(b) Plot the mean of the data on your graph. By eye fit a straight line passing through the mean to your plotted points. Use the line to estimate the average admission price in 1985.

Table 10.16

Year	1976	1978	1980	1982	1983	1984
Price (p)	73.0	93.7	141.3	177.4	189.8	194.3

(c) Table 10.17, extracted from the 1986 Edition of the Annual Abstract of Statistics, shows the purchasing power of the pound (based on the retail price index) in each of the years given in Table 10.16. The purchasing power in 1976 is taken to be 100 pence. Use this additional information to comment on the trend of the real cost of admission to cinemas over the period from 1976 to 1984.

Table 10.17

Year	Purchasing power (p)
1976	100
1978	80
1980	60
1982	49
1983	47
1984	45

☐ LEVEL 2

5 The annual production outputs, in thousands of kg, of a firm in the years from 1978 to 1985 are given in Table 10.18.
(a) Draw a line chart of the data.
(b) Calculate the four-year moving averages and plot them on your graph.
(c) Comment briefly on the general trend of the annual output.

Table 10.18

Year	Output ('000 kg)
1978	539
1979	494
1980	603
1981	586
1982	493
1983	402
1984	507
1985	474

6 Table 10.19, compiled from the records at a certain weather station, gives the number of days when snow fell in each of the years from 1975 to 1984.
(a) Draw a line chart of the data.
(b) Calculate the three-year moving averages and plot them on your graph. What do these points tell you about the general trend?

Table 10.19

Year	No. of snow days
1975	4
1976	5
1977	7
1978	5
1979	8
1980	11
1981	7
1982	12
1983	16
1984	10

(c) By eye fit a line to your plotted moving averages. Determine the slope of your fitted line and explain what information this provides about the frequency of snow at the station over the period from 1975 to 1984.

7 Table 10.20 gives the percentages of time lost through illness at a factory for each of twelve successive weeks.
(a) Plot the data on graph paper.
(b) Calculate the five-week moving averages, plot them on your graph and fit a line to them. Use this line to estimate (i) the value of the next moving average, and hence (ii) the percentage of time lost through illness in Week No. 13.

Table 10.20

Week No.	% time lost
1	11.7
2	11.6
3	10.4
4	9.1
5	9.1
6	8.2
7	9.2
8	7.0
9	6.3
10	6.6
11	6.7
12	5.5

8 Table 10.21 shows the turnovers, in thousands of pounds, of a large departmental store during periods of four consecutive months in the years from 1981 to 1984.
(a) Graph the data as a line chart. Explain why your chart indicates the presence of seasonal (cyclical) variation.
(b) Estimate the secular trend of the turnovers by plotting appropriate moving averages and fitting a straight line to them.
(c) Estimate the turnover during the first four months of 1985.
(d) Estimate the average seasonal effect during the first four months of a year.

Table 10.21

Year	Jan–Apr	May–Aug	Sep–Dec
1981	88	118	124
1982	119	150	154
1983	145	178	183
1984	177	209	215

9 Table 10.22 shows the number of colour television sets sold by a shop in each quarter of the years 1982, 1983, and 1984.
(a) Graph the data.
(b) On the same graph plot the values of moving averages appropriate for estimating the general trend of the number of sets sold. Estimate the number of sets sold by the shop in the first quarter of 1985.

Table 10.22

Year	1	2	3	4
1982	9	11	12	10
1983	11	12	14	11
1984	11	13	14	12

10 The observed values of a variable at 20 equal intervals of time are shown in Table 10.23.
(a) Plot the data on graph paper.
(b) Plot the values of moving averages appropriate for estimating the general trend on your graph, and fit a straight line to your plotted values. Estimate the value of the next moving average and, hence, the next value of the variable.

11 Table 10.24 gives the value of a certain price index for the months of January, April, July, and October from July 1982 to April 1985. Use moving averages to estimate the general trend in the value of the index. Obtain estimates of:
(a) the seasonal effect on the value of the index in January 1985,
(b) the average seasonal effect on the value of the index in a month of January.

Table 10.23

Time point	Observed value
1	30
2	28
3	30
4	32
5	41
6	45
7	29
8	25
9	26
10	30
11	39
12	43
13	25
14	21
15	23
16	29
17	33
18	37
19	22
20	17

12 The quarterly sales (in thousands of pounds) of a certain company over a period of three years are shown in Table 10.25
(a) Plot the data as a line chart on graph paper. Refer to the shape of your line chart to deduce that the sales are subject to seasonal variation.
(b) Calculate the moving averages that are appropriate for estimating the general trend, and plot them on your graph.
(c) Describing the methods you use estimate (i) the sales in each of the first two quarters of 1985, (ii) the average seasonal effect in the first quarter of a year.

Table 10.24

Year	January	April	July	October
1982	—	—	95.7	102.5
1983	100.4	97.6	100.5	108.4
1984	106.6	102.4	103.5	112.0
1985	107.3	105.5	—	—

Table 10.25

Quarter	1982	1983	1984
1	54	68	78
2	71	90	100
3	82	95	114
4	93	108	132

11
PROBABILITY

"There is one thing
certain, namely, that we can have
nothing certain; therefore
it is not certain that we can have
nothing certain."

Samuel Butler, First Principles',
Notebooks (1912)

11.1 Introduction

Descriptions such as "likely", "very unlikely", "probable" and "highly improbable", and so on, are often applied in everyday speech to events which are not predictable with absolute certainty. For example, a person seeing an overcast sky may well say (or think) "It is very likely that it will rain soon." In this chapter we shall consider ways of quantifying how likely it is that an event will occur, for events of a particular type.

A numerical assessment of the likelihood of an event occurring is called the **probability of the event**, or the **chance** that the event will occur.

Probabilities are measured on a scale from 0 to 1. An event which cannot possibly occur is assigned probability 0, and is referred to as an **impossible event**. For example, if today is a Tuesday then the event "tomorrow will be a Friday" is an impossible event; also, if an ordinary die (as used in many games) is thrown then the event "the score obtained will be $1\frac{1}{2}$" is an impossible event. Each of these two events will have probability 0. At the other end of the scale, an event which *must* occur is assigned probability 1, and is referred to as a **sure event** (or sometimes as a **certain event**, but this may be misleading because the word 'certain' is also used to mean 'particular'). "The sun will set today" is an example of a sure event having probability 1 of occurring. As another example consider the tossing of a coin. Ignoring the possibility that the coin will land on its edge the outcome of such a toss must be a head or a tail showing uppermost. Thus the event "head or tail" is a sure event and has probability 1.

● *EXERCISE 11.1*

1 State which of the following are possible values for the probability of an event: (a) 0 (b) 1 (c) $\frac{1}{2}$ (d) $-\frac{3}{4}$ (e) -1 (f) 1.3.
2 Give two further examples of impossible events.

3 Give two further examples of sure events.
4 Can you think of an event that has probability $\frac{1}{2}$ of occurring?

11.2 Equally Likely Outcomes: Elementary Events

Consider an action which may lead to any one of two or more possible outcomes occurring. The individual possible outcomes are often referred to as **elementary events** and are collectively referred to as the **sample space**. Suppose that there are good reasons for assuming that all the possible outcomes are equally likely to occur. Then, each outcome (or elementary event) has the same probability of occurring as every other possible outcome. Assigning probabilities to the elementary events (or outcomes) under this assumption is illustrated in the following examples.

○ EXAMPLE 11.1

What is the probability of throwing a score of 6 with an ordinary die?

When an ordinary die is thrown the possible scores that may be obtained are 1, 2, 3, 4, 5 and 6. The collection {1, 2, 3, 4, 5, 6} is the sample space in this case. Now if the die is well balanced and is thrown fairly then it is reasonable to suppose that all six possible scores are equally likely to occur. It follows that each possible score has the same probability of occurring as every other score. Since the total probability (that the score will be one of the six numbers) is 1, each score is assigned probability $\frac{1}{6}$. In particular, the probability that a 6 will be thrown is $\frac{1}{6}$.

The letter P is often used as an abbreviation for "probability of", in which case the last statement may be written more compactly as

$$P(\text{a score of } 6) = \tfrac{1}{6}.$$

○ EXAMPLE 11.2

Alan is one of five children in a family. If one of the five children is chosen at random, what is the probability that Alan will be chosen?

Choosing one object at random from a collection of objects ensures that every object in the collection has the same chance of being chosen. In this example this may be achieved by writing the names of the children on five slips of paper, placing the slips in a box and then drawing out one slip "blindly" from the box. (See also Section 7.2, p. 76.) This process ensures that each of the five children has the same chance of being chosen. With a total probability of 1 that one of the five children will be chosen, it follows that each child has probability $\frac{1}{5}$ of being chosen. In particular,

$$P(\text{Alan is chosen}) = \tfrac{1}{5}.$$

● EXERCISE 11.2

1 There are 5 books on a shelf; only one of the books is fictional. One book is chosen at random. Find the probability that it will be the fictional book.
2 The note pocket of a wallet contains three £5 notes and one £10 note. If one note is chosen at random, what is the probability that it will be the £10 note?
3 The ace of spades, the ace of hearts, the ace of clubs and the ace of diamonds are removed from an ordinary pack of playing cards. The four cards are then shuffled. What is the probability that the top card (of the four) is the ace of spades?
4 A bunch of flowers consists of one red, two white, and two blue flowers. One of the flowers is to be chosen at random from the five in the bunch. What is the probability that the chosen flower will be red?

11.3 Equally Likely Outcomes: Non-Elementary Events

It is often the case that an event of interest is one which occurs when any one of two or more of the possible outcomes occurs. When all the possible outcomes are equally likely to occur, the probability of such an event occurring is the ratio:

$$\frac{\text{number of outcomes for which event will occur}}{\text{total number of all possible outcomes}}$$

This result is illustrated in the examples that follow.

○ EXAMPLE 11.3

When a fair die is thrown, what is the probability

that the score obtained will be (1) an even number, (2) an exact multiple of 3?

Here there are 6 possible outcomes (scores), namely 1, 2, 3, 4, 5, and 6. Since the die is a fair one we may assume that all six possible outcomes are equally likely to occur (each with probability $\frac{1}{6}$).

(1) The even scores are 2, 4, and 6. Since the occurrence of any one of these three scores means that the event "an even score" occurs, using the above result we have:

$$P(\text{an even score}) = \tfrac{3}{6} = \tfrac{1}{2}.$$

(2) A score which is a multiple of 3 occurs only if the outcome is 3 or 6. Thus using the above result:

$$P(\text{score is a multiple of 3}) = \tfrac{2}{6} = \tfrac{1}{3}.$$

○ EXAMPLE 11.4

A class consists of 20 girls and 15 boys. If one pupil is chosen at random what is the probability that a girl will be chosen?

The outcome here (chosen pupil) may be any one of the 35 pupils in the class. Since the choice is made randomly we may assume that the 35 possible choices are equally likely. Since 20 of these outcomes correspond to a girl being chosen, we have:

$$P(\text{a girl is chosen}) = \tfrac{20}{35} = \tfrac{4}{7}.$$

○ EXAMPLE 11.5

A batch of 100 chicken eggs were classified according to size (1, 2, 3 or 4) and according to colour (brown or white). The results are given in Table 11.1. (For example, 16 eggs are brown of size 4, and 8 are white of size 3.)

(1) If one egg is chosen at random find the probabilities that it will be (a) brown of size 3, (b) of size 1, (c) white.

Table 11.1

Size	Colour Brown	White
1	12	10
2	18	12
3	20	8
4	16	4

(2) Given that an egg chosen at random was brown find the probability that it was of size 2.

(1) Since the choice is made randomly, the chosen egg is equally likely to be any one of the 100 eggs.

(a) Of the 100 eggs, we see from Table 11.1 that 20 are brown of size 3. It follows that:

$$P(\text{brown of size 3}) = \tfrac{20}{100} = 0.2.$$

(b) The number of size 1 eggs is:

$$12(\text{brown}) + 10(\text{white}) = 22,$$

and the required probability is:

$$P(\text{size 1}) = \tfrac{22}{100} = 0.22.$$

(c) The number of white eggs is seen to be:

$$10(\text{size 1}) + 12(\text{size 2}) + 8(\text{size 3}) + 4(\text{size 4}) = 34,$$

so that the probability of the chosen egg being white is:

$$P(\text{white}) = \tfrac{34}{100} = 0.34.$$

(2) The number of brown eggs is:

$$12(\text{size 1}) + 18(\text{size 2}) + 20(\text{size 3}) + 16(\text{size 4}) = 66.$$

Since we know that the chosen egg was brown and was chosen randomly, it is equally likely to be any one of the 66 brown eggs. Of the 66 brown eggs we see from the table that 18 are of size 2. It follows that the probability that the chosen brown egg is of size 2 is:

$$P(\text{brown egg is of size 2}) = \tfrac{18}{66} = \tfrac{3}{11} = 0.2727, (\text{to 4 d.p.}).$$

● EXERCISE 11.3

1 In a packet of 10 refills for a ballpoint pen, 5 of the refills contain blue ink, 3 contain red ink, and the remaining 2 contain green ink.
(a) If one of the refills is picked at random from the packet, calculate the probabilities that it will contain (i) blue ink, (ii) red ink, (iii) green ink, (iv) either red or green ink. (b) Calculate the probability that it will not contain green ink.

2 An ordinary pack of 52 playing cards consists of 13 hearts (red), 13 diamonds (red), 13 spades (black) and 13 clubs (black). If one card is dealt from a well-shuffled pack find the probabilities that it will be (a) spade, (b) a diamond, (c) a red card.

3 The results of an examination taken by the 30 pupils in a class were such that 5 had grade A, 7 had grade B, 10 had grade C, 5 had grade D, and the remaining 3 had grade E. If one of these 30 pupils is chosen at random find the probabilities that the chosen pupil had (a) grade A, (b) grade B, (c) grade C, (d) grade D, (e) grade E, (f) grade B or better (that is, grade A or B), (g) grade C or worse (that is, one of the grades C, D and E).

4 The 40 houses on an estate were classified as shown in Table 11.2.

(a) If one of the houses is chosen at random find the probabilities that it will be (i) a detached house having 4 bedrooms, (ii) a link house having 3 bedrooms, (iii) a semi-detached house, (iv) a house having 4 bedrooms, (v) a link house having 4 bedrooms.
(b) If one of the houses having 3 bedrooms is chosen at random find the probability that it will be a link house.
(c) Given that a randomly chosen house is semi-detached, find the probability that it has 3 bedrooms.

Table 11.2

No. of bedrooms	Type of house		
	Detached	Semi-detached	Link
2	1	2	5
3	3	6	11
4	5	7	0

5 Of the 100 tickets sold in a raffle, 40 were red, 30 were blue and 30 were green. The winning ticket is drawn at random.
(a) Find the probabilities that it is (i) red (ii) not blue.
(b) Every red ticket is even-numbered, while every blue ticket is odd-numbered; of the 30 green tickets, 10 are even-numbered and 20 are odd-numbered.
(i) Draw up a two-way table showing the classification of each ticket according to its colour and according to whether it is even-numbered or odd-numbered.
(ii) Find the probability that the winning ticket is green and even-numbered. (iii) Given that the winning ticket was green, find the probability that it was an even-numbered ticket.

11.4 Equally Likely Outcomes: Tree Diagrams

In the examples given so far it has been relatively easy to list the possible outcomes of some action. In more complicated situations, a **tree diagram** can be useful as indicated in the examples that follow.

○ EXAMPLE 11.6

A fair coin is tossed twice. Find the probability that exactly one head will be obtained.

Each of the possible outcomes of two tosses of a coin will need to take into account the separate outcomes of the individual tosses. The various possibilities are shown in the **tree diagram** in Figure 11.1, in which head and tail have been abbreviated as h and t respectively.

Figure 11.1

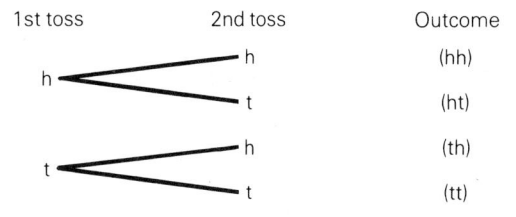

```
1st toss      2nd toss      Outcome
                    h         (hh)
      h  <
                    t         (ht)

                    h         (th)
      t  <
                    t         (tt)
```

We see that there are 4 possible outcomes for the two tosses, namely (hh), (ht), (th), and (tt). Note that (hh) represents the outcome "a head followed by a head", (ht) represents the outcome "a head followed by a tail", (th) represents the outcome "a tail followed by a head", and (tt) represents the outcome "a tail followed by a tail".

Since the coin is fair we may assume that the 4 possible outcomes are equally likely to occur (each with probability $\frac{1}{4}$). Of the 4 outcomes, those in which exactly one head was obtained are (ht) and (th). It follows that the probability of exactly one head is:

$$P(\text{exactly one head}) = \tfrac{2}{4} = \tfrac{1}{2}.$$

○ EXAMPLE 11.7

Three cards, numbered 1, 2 and 3 respectively, are shuffled and placed on a table in a row from left to right.
(1) Find the probabilities that (i) the numbers are 1, 2 and 3 in that order, (ii) the second number from the left is 2, (iii) the third number from the left is smaller than the second number from the left.
(2) Given that the first number is odd, find the probability that the second number is also odd.

The tree diagram to display the outcomes of the action here is given in Figure 11.2. The last column lists the possible outcomes. For example, the outcome (123) means that the cards, from left to right, are numbered 1, 2 and 3, in that order, while the outcome (132) means that the order of the numbers is 1 followed by 3 followed by 2. We see that there are 6 possible outcomes in all. Since the cards were shuffled beforehand we may assume that all 6 possible outcomes are equally likely to occur.

[143

Figure 11.2

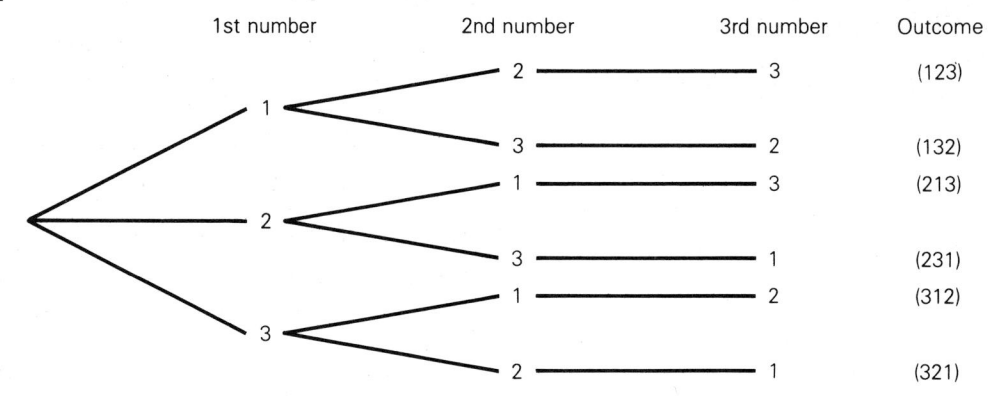

```
           1st number    2nd number    3rd number    Outcome
                             2 ─────────── 3         (123)
              1
                             3 ─────────── 2         (132)
                             1 ─────────── 3         (213)
   ─────── 2
                             3 ─────────── 1         (231)
                             1 ─────────── 2         (312)
              3
                             2 ─────────── 1         (321)
```

(1) (a) For the numbers to be 1, 2, 3 in that order, the outcome must be the one labelled (123). Hence the required probability is:

$$P(1, 2, 3 \text{ in that order}) = \tfrac{1}{6}.$$

(ii) The second number is 2 in each of the outcomes (123) and (321), so that:

$$P(\text{2nd number is 2}) = \tfrac{2}{6} = \tfrac{1}{3}.$$

(iii) The outcomes in which the third number is smaller than the second number are (132), (231), and (321), so that:

P(3rd number smaller than 2nd number $= \tfrac{3}{6} = \tfrac{1}{2}$.

(2) The outcomes in which the first number is odd are (123), (132), (312) and (321). We may assume that these four possibilities are equally likely to have occurred. Of these 4 outcomes the only ones in which the second number is odd are (132) and (312). Thus:

$$P(\text{2nd number is odd given 1st is odd}) = \tfrac{2}{4} = \tfrac{1}{2}.$$

● **EXERCISE 11.4**

1 With reference to Example 11.6 find the probability that the two tosses give different outcomes (that is, one toss gives a head and the other toss gives a tail).

2 With reference to Example 11.7 find the probabilities that:
(a) the first number is bigger than the third number,
(b) the sum of the first and third numbers is 4 or more,
(c) the product of the first two numbers is even.

3 A fair cubical die has two of its faces numbered 1, two numbered 2, and the remaining two faces numbered 3. The die is to be thrown twice.
(a) Draw a tree diagram to show the possible outcomes.
(b) Find the probabilities that in the two throws (i) the same score will be obtained, (ii) the score on the second throw will be greater than that on the first throw, (iii) the sum of the two scores will be 4.
(c) After the die had been thrown twice it was observed that the sum of the two scores was an even number. Find the probability that the two scores were 1 and 3, in any order.

4 Two cards are dealt one after another without replacement (that is, the first card is set aside after it has been dealt) from a shuffled pack of four cards which are numbered 1, 2, 3 and 4 respectively.
(a) Find the probabilities that the numbers on the dealt cards are (i) 2 and 4, in that order, (ii) such that one of them is exactly twice the other, (iii) such that their sum is 5 or more.
(b) Given that the number on the second card dealt is greater than that on the first card dealt, find the probability that the sum of the two numbers is 5 or more.

5 One box contains two balls, one of which is red and the other is white. Another box contains three balls, one of which is red, one is white, and the other is blue. One ball is drawn at random from the first box and then one ball is drawn at random from the second box. Find the probabilities that the two balls drawn are:
(a) both red,
(b) one white and one red (in any order),
(c) of the same colour,
(d) of different colours.

11.5 Mutually Exclusive Events

Two events that cannot both occur at the same time are said to be **mutually exclusive**, the occurrence of either excluding the possibility of the other also occurring.

Observe that any two elementary events (outcomes) are mutually exclusive. In Example 11.6 every two of the four possible outcomes (hh), (ht), (th), and (tt) are mutually exclusive. In Example 11.7 every two of the six possible outcomes (123), (132), (213), (231), (312), and (321) are mutually exclusive.

As an example of two non-elementary events that are mutually exclusive, suppose that a cubical die is thrown once. Then, the events "the score is an even number" and "the score is an odd number" are mutually exclusive, since the former occurs only if the score is 2, 4 or 6, while the latter occurs only if the score is 1, 3 or 5. Note, however, that the events "the score is a multiple of 3" and "the score is an even number" are not mutually exclusive since both occur if the score is 6.

☐☐ ADDITION RULE FOR MUTUALLY EXCLUSIVE EVENTS

For two mutually exclusive events the probability that either will occur is the sum of their individual probabilities of occurring.

Symbolically, if we denote the two mutually exclusive events by the letters A and B, then we have:

P(A or B will occur) = P(A will occur) + P(B will occur).

The following examples demonstrate this rule in cases when the outcomes are equally likely. (The rule is most useful when the outcomes are not equally likely, examples of which will be given in later sections.)

○ EXAMPLE 11.8

A fair die is thrown once. Find the probabilities that the score obtained will be (1) a multiple of 3, (2) an even number, (3) a multiple of 3 or an even number.

(1) The score will be a multiple of 3 if it is 3 or 6; these scores are mutually exclusive and each has probability $\frac{1}{6}$. By the addition rule:

P(score is a multiple of 3) = $\frac{1}{6} + \frac{1}{6} = \frac{1}{3}$,

exactly as obtained in Example 11.3(2).

(2) The score will be an even number if it is 2, 4 or 6; these scores are mutually exclusive since no two of them can occur together. Each of these three scores has probability $\frac{1}{6}$ of occurring, therefore:

P(score is an even number) = $\frac{1}{6} + \frac{1}{6} + \frac{1}{6} = \frac{1}{2}$,

exactly as obtained in Example 11.3(1).

(3) As indicated earlier, the events "the score is a multiple of 3" and "the score is an even number" are not mutually exclusive, so the addition rule should not be used here. The event "the score is a multiple of 3 or an even number" will occur if the score obtained is 2, 3, 4 or 6; these scores *are* mutually exclusive and each has probability $\frac{1}{6}$. It follows from the addition rule that:

P(score is a multiple of 3 or an even number) = $\frac{1}{6} + \frac{1}{6} + \frac{1}{6} + \frac{1}{6} = \frac{2}{3}$.

(Observe that the sum of the probabilities in (1) and (2) is $\frac{1}{3} + \frac{1}{2} = \frac{5}{6}$, which is not the correct answer to (3).)

○ EXAMPLE 11.9

With reference to Example 11.5, p. 142 find the probability that a randomly chosen egg will be brown of size 2 or white.

We first note that the events "brown of size 2" and "white" are mutually exclusive, so that:

P(brown of size 2 or white) = P(brown of size 2) + P(white).

From Table 11.1 given in Example 11.5, we find that:

P(brown of size 2) = $\frac{18}{100}$,

and:

P(white) = (10 + 12 + 8 + 4)/100 = 34/100.

Hence the required answer is:

$\frac{18}{100} + \frac{34}{100} = \frac{52}{100} = 0.52$.

● EXERCISE 11.5

Use the addition rule, whenever appropriate, to answer each of the following questions.

1 Referring to Example 11.5, p. 142 find the probabilities that a randomly chosen egg will be (a) white of size 3 or brown, (b) white or of size 3. Explain why the addition rule should not be used to answer (b).

2 Referring to Example 11.7, p. 143 find the probabilities that (a) the first number will be 1 or 3, (b) the second number will be 1 or 3.

3 With reference to Question 3, Exercise 11.4

find the probability that the sum of the two scores will be 3 or 4.

4 With reference to Question 4, Exercise 11.4 find the probabilities that (a) the second number dealt is 1 or 2, (b) the sum of the two numbers dealt is 3 or 4.

5 With reference to Question 5, Exercise 11.4 find the probabilities that (a) the two chosen balls will be of the same colour (check your answer with that your obtained for (c) of that question); (b) the number of red balls chosen will be 1 or 2.

11.6 Probability Tree Diagrams

Let us now consider an action which consists of two or more stages. We have already met some such actions. In Example 11.6 a coin was tossed twice. This may be regarded as a two-stage action, the first stage being the first toss of the coin and the second stage the second toss. In Example 11.7 three cards numbered 1, 2 and 3 were placed in a row. This may be regarded as a three-stage action, the stages corresponding to placing the first, second and third cards respectively.

☐ MULTIPLICATION RULE

When an action consists of successive stages, the probability of any one of the possible outcomes of the action is the product of the probabilities of the corresponding separate outcomes of the individual stages.

The following examples illustrate the use of this multiplication rule for probabilities.

○ EXAMPLE 11.10

A fair coin is tossed twice. Find the probability that exactly one head will be obtained.

From Example 11.6 the tree diagram for the possible outcomes is as shown in Figure 11.3. The individual probabilities associated with the various branches of the tree are also shown. Such a diagram is referred to as a **probability tree diagram**.

Since the coin is fair we know that in each toss the probability of a head (and of a tail) is $\frac{1}{2}$, and this is shown on each of the branches. The probability of any outcome of the two tosses is then obtained by multiplying the probabilities of the branches leading to that outcome, as shown on the extreme right of the diagram. Observe that these probabilities are identical to those we had in Example 11.6. The event "exactly one head" occurs if the outcome is (ht) or (th), which are mutually exclusive, and each has probability $\frac{1}{4}$. It follows that:

$$P(\text{exactly one head}) = \tfrac{1}{4} + \tfrac{1}{4} = \tfrac{1}{2},$$

agreeing with the answer we obtained in Example 11.6.

○ EXAMPLE 11.11

Reconsider Example 11.7 by regarding the action as consisting of three stages: 1st number, 2nd number, and 3rd number. The tree and the corresponding branch probabilities are shown in Figure 11.4.

Since the first number is equally likely to be anyone of 1, 2 and 3, each has probability $\frac{1}{3}$ as indicated along the branches for the first stage. What happens at the second stage depends upon what happened at the first stage. If the first number is 1 then the second number can only be 2 or 3; being equally likely, these numbers will each occur with probability $\frac{1}{2}$, as shown along the upper pair of branches at the second stage. Similar arguments lead to the other branch probabilities at the second

Figure 11.3

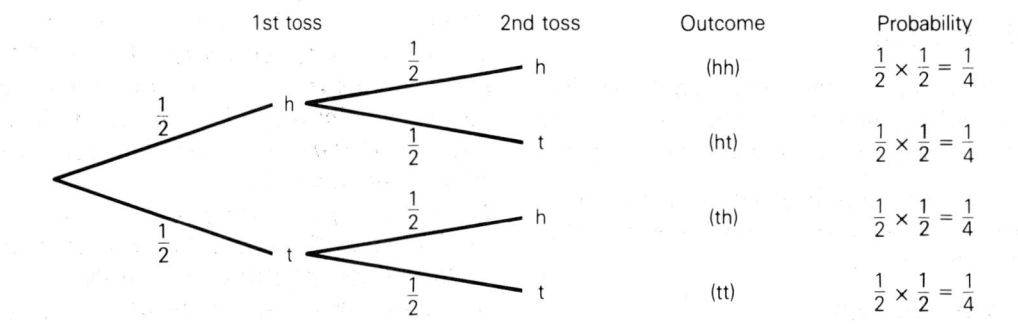

1st toss	2nd toss	Outcome	Probability
h ($\frac{1}{2}$)	h ($\frac{1}{2}$)	(hh)	$\frac{1}{2} \times \frac{1}{2} = \frac{1}{4}$
h ($\frac{1}{2}$)	t ($\frac{1}{2}$)	(ht)	$\frac{1}{2} \times \frac{1}{2} = \frac{1}{4}$
t ($\frac{1}{2}$)	h ($\frac{1}{2}$)	(th)	$\frac{1}{2} \times \frac{1}{2} = \frac{1}{4}$
t ($\frac{1}{2}$)	t ($\frac{1}{2}$)	(tt)	$\frac{1}{2} \times \frac{1}{2} = \frac{1}{4}$

Figure 11.4

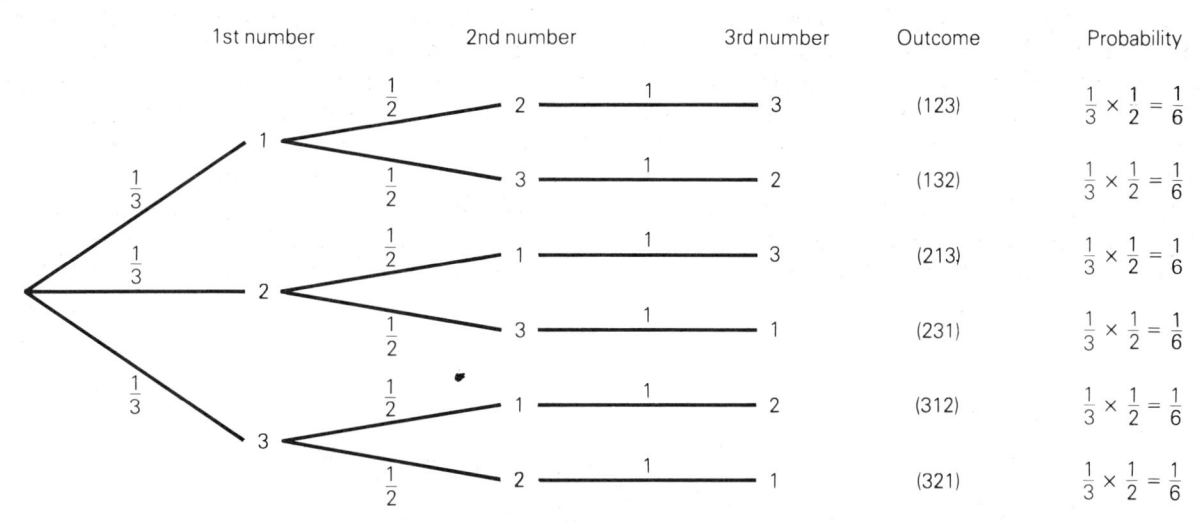

1st number	2nd number	3rd number	Outcome	Probability
	$\frac{1}{2}$ → 2	1 → 3	(123)	$\frac{1}{3} \times \frac{1}{2} = \frac{1}{6}$
$\frac{1}{3}$ → 1	$\frac{1}{2}$ → 3	1 → 2	(132)	$\frac{1}{3} \times \frac{1}{2} = \frac{1}{6}$
$\frac{1}{3}$ → 2	$\frac{1}{2}$ → 1	1 → 3	(213)	$\frac{1}{3} \times \frac{1}{2} = \frac{1}{6}$
	$\frac{1}{2}$ → 3	1 → 1	(231)	$\frac{1}{3} \times \frac{1}{2} = \frac{1}{6}$
$\frac{1}{3}$ → 3	$\frac{1}{2}$ → 1	1 → 2	(312)	$\frac{1}{3} \times \frac{1}{2} = \frac{1}{6}$
	$\frac{1}{2}$ → 2	1 → 1	(321)	$\frac{1}{3} \times \frac{1}{2} = \frac{1}{6}$

Figure 11.5

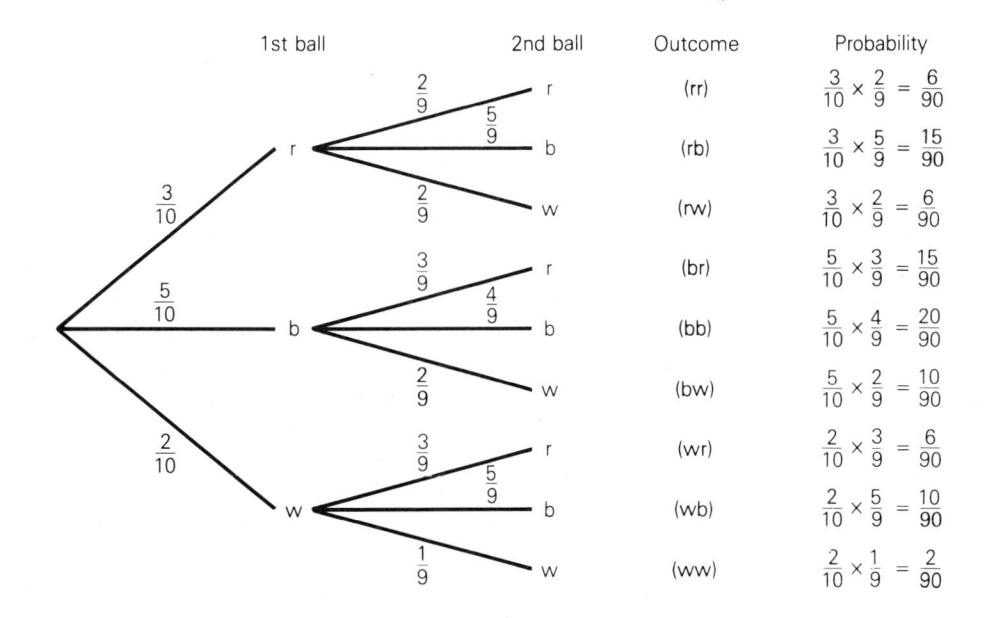

1st ball	2nd ball	Outcome	Probability
	$\frac{2}{9}$ → r	(rr)	$\frac{3}{10} \times \frac{2}{9} = \frac{6}{90}$
$\frac{3}{10}$ → r	$\frac{5}{9}$ → b	(rb)	$\frac{3}{10} \times \frac{5}{9} = \frac{15}{90}$
	$\frac{2}{9}$ → w	(rw)	$\frac{3}{10} \times \frac{2}{9} = \frac{6}{90}$
	$\frac{3}{9}$ → r	(br)	$\frac{5}{10} \times \frac{3}{9} = \frac{15}{90}$
$\frac{5}{10}$ → b	$\frac{4}{9}$ → b	(bb)	$\frac{5}{10} \times \frac{4}{9} = \frac{20}{90}$
	$\frac{2}{9}$ → w	(bw)	$\frac{5}{10} \times \frac{2}{9} = \frac{10}{90}$
	$\frac{3}{9}$ → r	(wr)	$\frac{2}{10} \times \frac{3}{9} = \frac{6}{90}$
$\frac{2}{10}$ → w	$\frac{5}{9}$ → b	(wb)	$\frac{2}{10} \times \frac{5}{9} = \frac{10}{90}$
	$\frac{1}{9}$ → w	(ww)	$\frac{2}{10} \times \frac{1}{9} = \frac{2}{90}$

stage. At the third stage there is only one card left. If the first two numbers are 1 and 2 then the third number has to be 3, with probability 1 as indicated along the top branch at the third stage.

Multiplying together the probabilities along any route of branches gives the probability of the associated outcome. The possible outcomes and their probabilities are shown on the extreme right of the diagram. As assumed in Example 11.7, we see that all six outcomes are equally likely, each having probability $\frac{1}{6}$.

The real advantage of the addition and multiplication rules is that they are equally valid when the action outcomes are not all equally likely to occur, as in the following example.

○ **EXAMPLE 11.12**

Two balls are drawn at random one after the other without replacement from a bag containing 3 red balls, 5 blue balls, and 2 white balls. Find the probabilities that (1) both balls are red, (2) the two balls are of the same colour, (3) the second ball drawn is red.

The probability tree diagram is given in Figure 11.5. In this diagram r denotes a red ball, b denotes a blue ball and w denotes a white ball.

The probabilities indicated along the branches at the first stage are obtained from the fact that one ball is chosen at random from 10 balls of which 3 are red, 5 are blue and 2 are white. At the second stage we need to take account of what happened at the first stage. If the first ball is red then, since the draws are made *without replacement*, the second draw is to be made from a bag containing 9 balls of which 2 are red, 5 are blue and 2 are white. The probability of a red is now $\frac{2}{9}$, that of a blue is $\frac{5}{9}$, and that of a white is $\frac{2}{9}$, as indicated along the uppermost trio of branches at the second stage. The probabilities along the other branches at the second stage are obtained similarly. The probability of a possible outcome is then obtained by multiplying together the probabilities along the branches leading to that outcome.

Observe that here we have an instance where the possible outcomes are not equally likely to occur. This is evident from the probabilities of the outcomes as given in the extreme right column in the diagram.

Having constructed a probability tree it is always advisable to carry out the following checks.

(a) Check that the sum of the probabilities along branches from a common point is equal to 1, as it must be since all possibilities from the common point are being considered. At the first stage in

Courting Disaster

Why is it apparently so difficult to find a suitable person with whom to date? Probably because we expect too much. Make a list of the ten qualities that you would most like to find in your boy/girl-friend. You might like to start with:

Likes me.
Doesn't have a regular boy/girl-friend already.
etc.

When you have your list of ten qualities, give each one a value between zero and one to indicate how likely it is that a person you will meet will satisfy that characteristic. Of all the people I meet, perhaps a fifth might like me, and perhaps one in two would be unattached, and so on. Multiplying these probabilities together to work out how likely you are to meet someone with all these characteristics can be rather discouraging. Suppose I get the answer 1/300 000. This means that if I meet one new person of the right sex each day, I will only meet a "dream person" about once every 800 years on average!

One adjustment that I could make to my list would by to compromise on those items which have very low probabilities. Perhaps "over 190 cm" is a bit over-ambitious. Again, a more careful look at my list may show that some of the items are not independent. Once I have found a 'wealthy' person, I may well have found one who is also 'ambitious'. I could therefore adjust the probability of 'ambition' from 1/3 to 2/3, say, conditional upon 'wealth'. Unfortunately, my list may also contain incompatibles. 'Ambition' and 'sensitivity' do not go together that often. Perhaps I should choose my priority and hope that the other characteristic may develop in time. In any event, there is plenty of scope for shortening such a demanding "shopping-list".

Having adjusted the list to improve the chance of finding someone who fits my demands, I can increase the probabilities by looking in the right place. For example, if I want someone athletic, the probability of meeting an appropriate person will be increased if I spend more time at the local sports centre than at the cinema.

If you should decide to adopt such dating strategies yourself, do not be misled by the gamblers' fallacy. Meeting a lot of "almosts" (people with some, but not all, of the characteristics), will not increase the chance of the next person you meet being a "person of your dreams". You will be in for a great many disappointments if you expect this to be the case.

our example the branch probabilities are $\frac{3}{10}$, $\frac{5}{10}$ and $\frac{2}{10}$, which do add up to 1. The probabilities along the three branches at the second stage, following a red ball at the first stage, are $\frac{2}{9}$, $\frac{5}{9}$ and $\frac{2}{9}$, which also add up to 1. The same will be found to be true for the other two trios of branches at the second stage.

(b) Check that the sum of the probabilities of all the possible outcomes is equal to 1. This is readily verified in our example from the column of probabilities given on the extreme right of Figure 11.5.

(1) For both balls to be red the outcome must be (rr). From Figure 11.5 we see that:

P(both red) = $\frac{6}{90}$ = $\frac{1}{15}$ = 0.0667   (to 4 d.p.).

(2) The two balls are of the same colour if the outcome is one of (rr), (bb) and (ww). Since these outcomes are mutually exclusive, the required probability is the sum of the probabilities of these outcomes. From Figure 11.5 we find that:

P(same colour) = P(rr) + P(bb) + P(ww)
= $\frac{6}{90}$ + $\frac{20}{90}$ + $\frac{2}{90}$ = $\frac{28}{90}$ = 0.3111
(to 4 d.p.).

(3) The second ball is red if the outcome is one of (rr), (br) and (wr). Hence:

P(2nd ball is red) = P(rr) + P(br) + P(wr)
= $\frac{6}{90}$ + $\frac{15}{90}$ + $\frac{6}{90}$
= $\frac{27}{90}$ = 0.3.

● **EXERCISE 11.6**

1 Suppose that for a particular married couple each child born is equally likely to be a boy or a girl. Draw a probability tree diagram to show the various possible outcomes and their probabilities for the first three children born. Are the outcomes equally likely? Find the probabilities that the first three children (a) are all girls, (b) include at least two girls, (c) consist of more boys than girls, (d) are such that the youngest child is a girl, (e) are such that the sexes alternate according to order of birth.

2 A box contains 2 white and 2 black balls. The balls are withdrawn from the box one after another at random without replacement. Find the probability that in the order they are drawn the balls alternate in colour.

3 From a box containing 6 white balls and 2 black balls, three balls are drawn at random. Find the probability of drawing 2 white balls and 1 black ball (a) when the balls are drawn without replacement, (b) when the balls are drawn with replacement (a drawn ball being replaced in the box before the next draw is made).

4 It is known that the probability of a manufactured item being defective is 0.1. If 3 such items are chosen at random from the total output find the probabilities that (a) none of the chosen items is defective, (b) exactly one of the chosen items is defective, (c) 1 or 2 of the chosen items are defective.

5 Suppose that the probability of a man aged 50 being alive when aged 60 is 0.7, and that the probability of a woman aged 50 being alive when aged 60 is 0.9. Find the probabilities that in 10 years time a husband and wife both aged 50 will (a) both be alive, (b) both be dead.

11.7 Further Examples

One big disadvantage with the probability tree diagram is that it becomes rather large when several stages are involved and/or the number of possibilities at one or more of the stages is large. In many cases it is sufficient to draw only that part of the probability tree diagram which is relevant to the event under consideration.

○ **EXAMPLE 11.13**

Suppose that in Example 11.12 we only required the probability that the second ball drawn is red. To find this we only need that part of the full probability tree diagram leading to the second ball being red; that is, to the outcomes (rr), (br) and (wr), as shown in Figure 11.6.

The required probability is the sum of the three probabilities listed, as given in the solution to Example 11.12(3).

○ **EXAMPLE 11.14**

Three cards are dealt at random one after another from an ordinary pack of playing cards. Find the probability that the first card dealt will be an ace, the second will be an ace or a king, and the third will be a king or a queen.

A pack of cards consists of four aces, four kings, four queens and 40 other cards. The possible ways in which the event of interest can occur are shown in the Figure 11.7, in which A represents an ace, K represents a king and Q represents a queen. Summing the probabilities the required answer is:

Figure 11.6

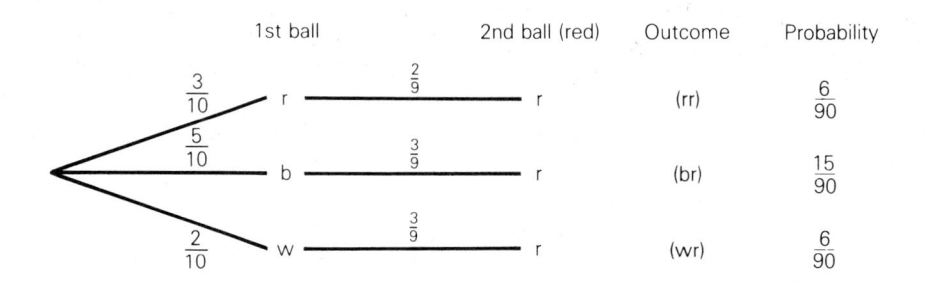

```
                        2/9
          3/10  r ─────────── r      (rr)       6/90
         ╱
        ╱  5/10      3/9
       ─────── b ─────────── r      (br)       15/90
        ╲
         ╲ 2/10      3/9
          ╲    w ─────────── r      (wr)       6/90
```

Figure 11.7

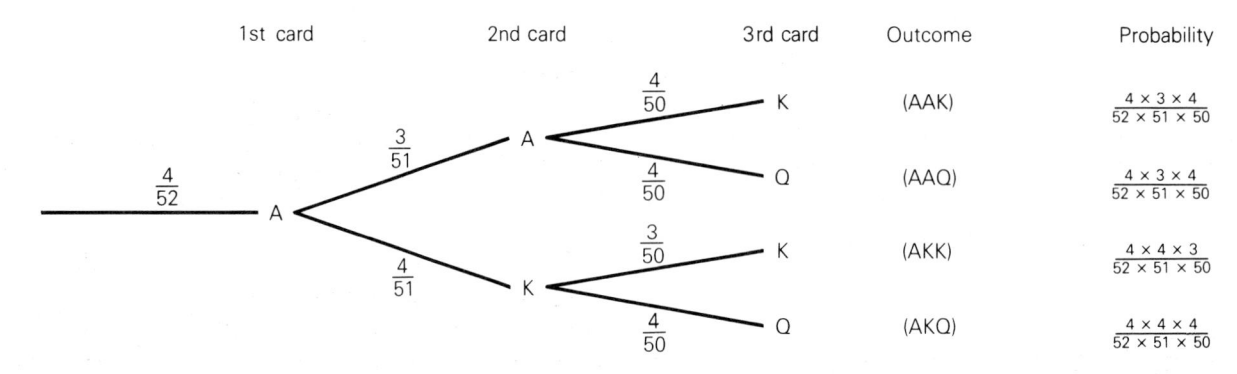

```
                              4/50
                        ┌──────── K     (AAK)    (4×3×4)/(52×51×50)
                  3/51  │
              ┌──── A ──┤ 4/50
              │         └──────── Q     (AAQ)    (4×3×4)/(52×51×50)
   4/52       │
  ──── A ─────┤         
              │         ┌ 3/50
              │ 4/51    │
              └──── K ──┤────── K     (AKK)    (4×4×3)/(52×51×50)
                        │
                        └ 4/50
                          ────── Q     (AKQ)    (4×4×4)/(52×51×50)
```

$$\frac{(4\times3\times4)+(4\times3\times4)+(4\times4\times3)+(4\times4\times4)}{52\times51\times50}$$

$$=\frac{208}{52\times51\times50}=0.0016 \text{ (to 4 d.p.)}$$

(to 6 d.p.).

(Observe that the complete tree diagram, which would have to take account of the four different types of cards (ace, king, queen and 'others'), would involve $4\times4\times4=64$ distinct routes to cover all possible outcomes.)

● **EXERCISE 11.7**

1 A Youth Club committee has 10 members, consisting of 6 girls and 4 boys. Four members of the committee are to be chosen at random to represent the Club at a Youth Festival. Find the probabilities that the four chosen will be (a) all boys, (b) all girls, (c) all of the same sex.

2 Of the 20 gums in a bag, 10 are lemon-flavoured, 6 are orange-flavoured, and 4 are strawberry-flavoured. If 3 gums are chosen at random from the bag without replacement find, to two decimal places, the probabilities that (a) all 3 will be of the same flavour, (b) the first will be lemon-flavoured and the third strawberry-flavoured.

3 A fair die is thrown twice. Find the probabilities that (a) both scores will be even numbers, (b) in the order obtained, the two scores will be increasing consecutive integers (for example, 1 2 or 2 3, and so on), (c) exactly one 6 will be obtained.

4 A fair die is thrown three times. Find the probabilities that (a) all three scores will be even numbers, (b) in the order obtained, the scores will be increasing consecutive integers (for example, 1 2 3 or 2 3 4, and so on), (c) exactly one 6 will be obtained.

5 A fair coin is tossed once. If a head is tossed then a fair die is thrown once and the number on the die is recorded as the score. If the toss results in a tail then a fair die is thrown twice and the sum of the two numbers obtained is recorded as the score. Find the probabilities that the recorded score will be (a) 2 (b) 3 (c) 4.

6 The four cards in a pack are numbered 1, 2, 3, and 4 respectively. Two of the cards are chosen at random. Find the probability that in the order in which they are drawn the two cards are increasing consecutive integers (a) when the cards are chosen without replacement, (b) when the cards are chosen with replacement.

11.8 Relative Frequency Interpretation of Probability

So far we have been concerned solely with determining the probabilities of events when the possible outcomes of an action (or of each stage of a multistage action) are equally likely to occur. This is so in many games of chance involving coins, dice, or cards, and in the case of randomly choosing a sample of objects from a collection.

However, there are events whose probabilities cannot be determined using the methods already described. To take a particular example, what is the probability that a drawing pin dropped onto a table will come to rest with its point facing upwards? Just like the tossing of a coin there are two possible outcomes of the action here, namely that the point is upwards or the point is in contact with the table. But there is an important difference. Unlike a coin a drawing pin does not have physical symmetry to justify the assumption that the two outcomes are equally likely to occur. Some other examples of actions having two possible outcomes which are not generally equally likely to occur include:

(1) a surgical operation (successful or not),
(2) treatment for a certain ailment (successful or not),
(3) germination of a planted seed (germinates or does not germinate),
(4) birth of a child (boy or girl).

Consider some arbitrary event associated with an action whose outcomes cannot be assumed to be equally likely to occur. Suppose that in a very large number of performances of the action the relative frequency of occurrence of the event (that is, the proportion of the performances when the event occurred) was 0.2. We then take the value 0.2 as the probability of the event occurring in any subsequent performance of the action. For example, birth statistics over several years show that about 51% of all births are boys. Thus we can take the probability that an unborn child is a boy to be about 0.51. (In Question 1 of Exercise 11.6 you were told to assume that a boy and a girl are equally likely; this is equivalent to the assumption that the probability of an unborn child being a boy is 0.5, which is slightly less than the value 0.51 (obtained from records).

One of the problems facing a life insurance company is that of determining the expected length of a person's life in order to fix the level of premiums to be paid by that person. To do so, the company makes use of mortality tables. These give relative frequencies which are equivalent to the probabilities that people of given ages and sexes will survive a given number of years.

Another example of the use of a relative frequency as a probability is any estimate of the success rate of a surgical operation or treatment which is based on the outcomes of past operations or treatments of the same type.

Let us now return to our drawing pin experiment and try to find the probability that when dropped onto a table the point will be upright. To do this a particular drawing pin was dropped onto a table 100 times and on each occasion a record was make of whether or not the point was upright. A table was then compiled of the numbers, and hence the relative frequencies, of occurrences of "point upright" in the 1,2,3,... 100 performances of the experiment. A summary of the results obtained is given in Table 11.3 and is displayed graphically in Figure 11.8.

Table 11.3 *Relative Frequencies of "Point Upright"*

No. of performances	No. upright	Relative frequency
10	1	0.1
20	5	0.25
30	9	0.333
40	13	0.325
50	15	0.3
60	20	0.333
70	22	0.314
80	24	0.3
90	28	0.311
100	30	0.3

Figure 11.8 *Plot of the Data in Table 11.3*

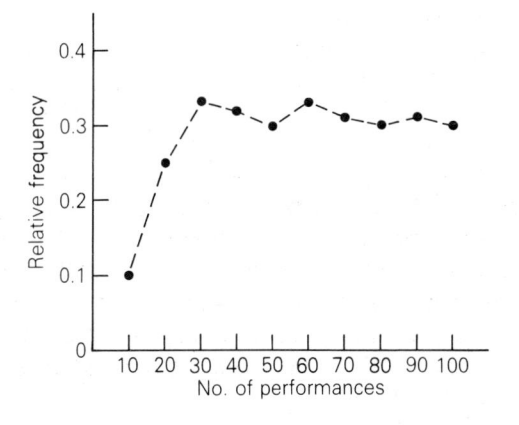

It is evident from Figure 11.8 that the relative frequencies increase fairly substantially for up to the first 30 performances. They then settle down to vary only slightly, eventually seeming to stabilize at a value which is about 0.3. The experiment suggests that the probability of this particular drawing pin landing with its point upright is about 0.3. This stabilizing phenomenon of the relative frequency has been confirmed experimentally for several types of events associated with an action that can be performed indefinitely under identical conditions.

You may like to carry out the drawing pin experiment yourself, but as your drawing pin may differ in its shape from the one used in the above experiment your estimate of the probability that it will land with its point upright may be very different from the value 0.3 obtained above. Alternatively, or additionally, you could carry out some of the experiments listed as possible projects at the end of this chapter.

Recapping, when we say that the probability of an event is P we mean that the event will occur in a proportion P of a very large number of performances of the action under consideration. If the action is to be performed n times then we refer to the product nP as being the **expected frequency** of the event in those n performances.

For example, if a fair coin is tossed we know that the probability of a head is $\frac{1}{2}$. If the coin is to be tossed 50 times the expected number (frequency) of heads is $50 \times \frac{1}{2} = 25$. This does not mean that 25 heads will be obtained in the 50 tosses. Different sets of 50 tosses will give varying numbers of heads. But for a large enough number of sets of 50 tosses the mean number of heads per set would be about 25, the expected frequency as defined here.

As another example, if a fair die is to be thrown 50 times then the expected frequency of a score of 6 is $50 \times \frac{1}{6} = 8\frac{1}{3}$. In this case it is clear that the number of 6's that will occur cannot possibly be $8\frac{1}{3}$, the expected frequency (or number). However, in a very large number of sets of 50 throws of the die the mean number of 6's per set would be approximately equal to $8\frac{1}{3}$.

○ EXAMPLE 11.15

Two balls are to be drawn at random one after the other without replacement from a bag that contains 3 red balls, 5 blue balls, and 2 white balls. If this action is carried out a total of 30 times, in each of which the bag's contents are as stated above, find the expected number of times that the two balls drawn will (1) both be red, (2) be the same colour.

(1) In Example 11.12(1) we found that:

$$P(\text{both will be red}) = \frac{1}{15}.$$

In 30 performances of the action the expected number in which both balls will be red is $30 \times \frac{1}{15} = 2$.

(2) In Example 11.12(2) we found that:

$$P(\text{same colour}) = \frac{28}{90}.$$

In 30 performances of the action the expected number in which the two balls will be of the same colour is $30 \times \frac{28}{90} = 9\frac{1}{3}$.

● EXERCISE 11.8

1 An analysis of the results of the matches played by teams in a certain competition during a season showed that a total of 942 matches were played. Of these, 512 were home wins, 246 were away wins, and the remaining 184 matches were drawn.
(a) Estimate to three decimal places the probabilities that in a match next season in this competition the result will be (i) a home win, (ii) an away win, (iii) a draw.
(b) Of the 1884 matches played over two seasons what are the expected numbers of matches that are (i) home wins, (ii) away wins, (iii) draws.

2 The probability that a manufactured item is defective is stated to be 0.1.
(a) Suggest how this value may have been obtained. The items are packaged into cartons, each carton containing 120 items. What is the expected number of defectives per carton?
(c) How would you explain the meaning of your answer to someone who was unfamiliar with probability and expected number?

3 Two dice were thrown together fifty times and after each throw a record was made of the sum of the two scores on the dice. It was found that in the 50 throws the sum of the scores was 5 in 6 of them, 7 in 10 of them, and 10 in 4 of them.
(a) If the dice are to be thrown together once more, estimate the probabilities that the sum of the two scores will be (i) 5 (ii) 7 (iii) 10.
(b) Assuming that the two dice are fair, calculate the theoretical probabilities that in one throw the sum of the scores will be (i) 5 (ii) 7 (iii) 10.

4 A bag contains 8 black balls and 2 white balls.
(a) If two balls are drawn at random without replacement from the bag find the probabilities that (i) both will be black, (ii) they will be of the same colour. (b) Suppose that this action is to be carried out 40 times, with the contents of the bag being as described above on each occasion. Find the expected numbers of times that the two chosen balls will (i) both be black, (ii) be the same colour.

*Projects

1 Conduct at least 100 performances of an experiment similar to the drawing-pin experiment described in Section 11.8 to demonstrate the long-run stability of the relative frequency of the occurrence of an event. Some suggestions which do not involve any technical knowledge (other than that contained in this book) or any specialist equipment are given below. Where appropriate, the worked example giving the theoretical probability is stated.
(a) Scoring 6 in one throw of a die (Example 11.1).
(b) Getting exactly one head in two tosses of a coin. (Example 11.6 or 11.10).
(c) Each of the events of Example 11.7.
(d) Ending up with a girl when choosing one pupil at random from a class of 20 girls and 15 boys (Example 11.4). (This action can easily be simulated using a pack of ordinary playing cards. Remove any 20 red cards (to represent the girls) and any 15 black cards (to represent the boys). Shuffle this pack of 35 cards and then randomly deal one card from it. Note the colour of the card, recording 'girl' if it is red and 'boy' if it is black. Replace the card in the reduced pack and repeat the entire process until you have generated sufficient results for the purpose of this project.)
(e) Each of the events of Example 11.5(1) (The action here may be simulated by taking pairs of random digits from a table (such as Table 7.1, p. 77) and treating the occurrence of any one of 00 to 11 as having chosen a brown egg of size 1, any one of 12 to 21 as having chosen a white egg of size 1, and so on, ending up with the occurrence of any one of 96, 97, 98, and 99 as having chosen a white egg of size 4.)

2 A person is asked to choose a number from 1 to 9 inclusive. Assuming that the person makes a random choice calculate the probabilities that the chosen number will be (a) 7 (b) odd.

It has been claimed that people do not in fact make random choices when asked to give a number from 1 to 9. Try it out on at least 100 persons and compile a frequency distribution of the responses you obtain. Which number (or numbers) was the most popular choice? Also obtain estimates of the probabilities of the events in (a) and (b).

3 Published data, such as that contained in H.M.S.O. publications, provide a valuable source for estimating the probabilities of a wide variety of events. Use some published data for this purpose. For example: (a) What is the probability that a primary school chosen at random from those in your county (or country) has fewer than 150 pupils? (b) What is the probability that a household chosen at random from the households in your county (or country) has (i) a colour T.V., (ii) at least 3 children attending school?

Review Exercises Chapter 11

☐ LEVEL 1

1 A pupil claims that the chance of a man landing on Mars sometime this century is $-\frac{1}{2}$. Do you believe this? Justify your answer.

2 A particular pupil feels that her probability of passing a Statistics examination is $\frac{3}{4}$ and that her probability of failing is $\frac{1}{2}$. Explain why her assessments cannot possibly be correct.

3 Give any one example of an event whose probability of occurring is $\frac{1}{3}$.

4 A teacher opens a new box of white chalk sticks. There are 150 sticks in the box; 12 of them have broken into two pieces but look to be perfect. If the teacher chooses one stick at random from the box find the probability that it is a broken stick.

5 In a magnetic fishing game at a fairground a small pool of water contains 25 (metal) fish, of which 5 are marked as prize winners. A person plays the game and catches one of the fish. What is the probability that this person will win a prize?

6 A die in the form of a tetrahedron (a solid with four triangular faces) has its four faces numbered 1, 2, 3, and 4 respectively. When the die is rolled on a table the score obtained is the number on the face which is in contact with the table. You may assume that the die is perfectly symmetrical so that all four possible scores are equally likely to occur. In one throw of this die find the probabilities that the score obtained will be (a) 4 (b) an even number.

7 According to mortality tables, of every 1000 men who are aged 50 years, only 869 will survive to the age of 60 years. What is the probability that a man aged 50 years will live to be 60 years old?

8 When asked to find the probability that when a fair coin is tossed twice 2 heads will be obtained, a particular pupil produced the following solution: "If a fair coin is tossed twice the three possible outcomes are 0 heads, 1 head, and 2 heads. It follows that the probability of 2 heads is $\frac{1}{3}$." Where has the pupil gone wrong? Give the correct solution.

9 Two balls are drawn at random without replacement from a bag containing 2 white balls and 3 black balls.
(a) Find the probability that both balls drawn will be white.
(b) If this action is to be performed 50 times, what is the expected number of times that 2 white balls will be drawn?

10 The 45 pupils in a class were classified according to sex and according to hair colour. The results obtained are given in Table 11.4.
(a) Calculate the probabilities that a pupil chosen at random from the class will be (i) male with light coloured hair, (ii) female, (iii) one having reddish coloured hair, (iv) either a male having dark coloured hair or a female.
(b) A pupil chosen at random turned out to be female. Find the probability that she had dark coloured hair.
(c) If the action of choosing a pupil at random from the 45 pupils in the class is carried out 50 times what is the expected frequency of the chosen pupil having light coloured hair?

Table 11.4

Sex	Hair colour		
	Reddish	Light	Dark
Male	1	12	7
Female	2	15	8

☐ LEVEL 2

11 I have three flower beds in my garden. The first bed has 40 flowers in bloom and 10 of them are red. The second bed has 30 flowers in bloom and 12 of them are red. The third bed has 50 flowers in bloom and 25 of them are red. If I choose one of the three beds at random and then randomly choose a flower from it find the probabilities that I will choose (a) the first bed and a red flower from it, (b) a red flower.

12 Bag A contains 6 red balls and 4 white balls.
(a) If two balls are to be drawn at random without replacement from this bag find the probability that the second ball drawn will be red.
(b) Bag B contains 2 red and 8 white balls. If one ball is drawn at random from bag A and one ball is drawn at random from bag B find the probability that the balls are of different colours.

13 A fair coin is tossed three times.
(a) Find the probabilities that (i) all three outcomes are the same (that is, 3 heads or 3 tails), (ii) there are exactly 2 heads.
(b) What is the expected frequency of exactly 2 heads occurring if the action of tossing a fair coin three times is carried out 100 times?

14 A bag contains 8 red balls and 4 white balls.
(a) One ball is to be drawn at random from the box. Find the probabilities that it will be (i) red, (ii) white.
(b) Two balls are to be drawn at random without replacement from the bag. Find the probabilities that (i) both will be red, (ii) they will be the same colour, (iii) they will not both be white.
(c) Suppose that some extra red balls are placed in the bag. How many red balls were added if the probability of drawing a red ball is now $\frac{3}{4}$?

15 Two cards are drawn at random without replacement from a pack of 10 cards which are numbered from 1 to 10, respectively. Calculate the probabilities that (a) the number on the first card drawn is 5, (b) the number on the first card is even, (c) the number on the second card is even, (d) the number on the second card is exactly three times the number on the first card.

16 Repeat Question 15 but this time the two cards are drawn with replacement.

17 In two throws of a fair cubical die find the probabilities that (a) the two scores are the same or add to 9, (b) the two scores are the same or add to 10.

18 In a certain game a player throws a fair die. If a 6 is not thrown the player's score is the number showing on the die. If a 6 is thrown the player has one more throw and scores 6 plus the number showing on the die in the second throw. Find the probabilities that the player's score in one turn will be (a) less than 6, (b) 10 or more.

19 A box contains 5 torch batteries but 2 of them are dead. The batteries are withdrawn from the box one at a time and tested. This is done until both dead batteries have been identified. Find the probabilities that the number of batteries tested is (a) 2 (b) 3 (c) 4.

20 In a football competition the probability that a match will result in a home win is 0.7 and in a draw is 0.1.
(a) What is the probability that a match will result in any away win?
(b) If a total of 64 matches are played in the competition find the expected number that will result in (i) home wins, (ii) draws, (iii) away wins.

12

PICTOGRAMS, MISLEADING DIAGRAMS, AMBIGUOUS STATEMENTS AND MISINTERPRETATIONS

"What sort of a data-bird are you — an ostrich, a blinker or a statistickler?"

Figure 12.1 *Enhanced Pictorial Representations of Data*

(a) BREAKDOWN OF AN AVERAGE GALLON OF 4 STAR PETROL

12.1 Enhanced Diagrams

A newspaper or magazine article reporting on some statistical data will often have an accompanying diagram which is more artistic and pictorial than the ones we have introduced in earlier chapters, the aim being to attract a reader's attention. Magazines and television programmes will enhance the diagram further by using colour. Figures 12.1a and b are two examples of enhanced pictorial representations of data that appeared in newspapers in early 1986. These pictures are, respectively, enhanced pictorial representations of a component bar diagram (see Section 2.2, p. 8) and pie diagrams (see Section 2.2, p. 9 and Section 2.4, p. 12).

Another pictorial form for representing data is one in which appropriate symbols are used. Such a diagram is referred to as a **pictogram** or an **ideograph**. The purpose of a pictogram is to provide a simple diagrammatic representation to convey fairly

Petrol Station Gross Margin — 6%
Oil Company Costs — 8%
Cost of Product — 35%
UK Government Tax — 51%

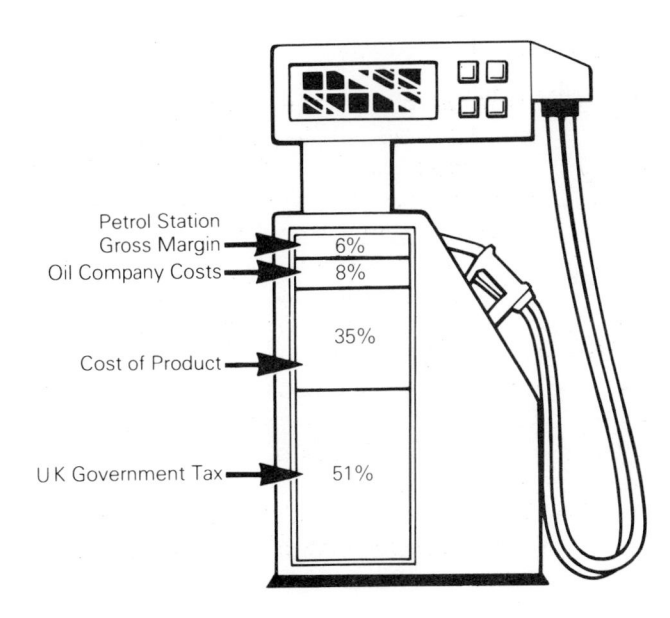

[155

Figure 12.1 (*continued*)

(b) DISTRIBUTION OF BLENDED SCOTCH WHISKY SALES (1985)

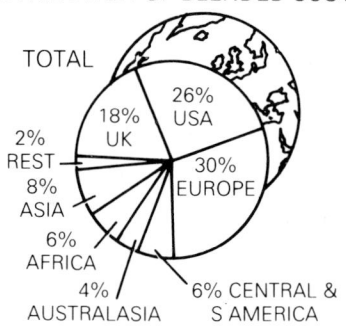

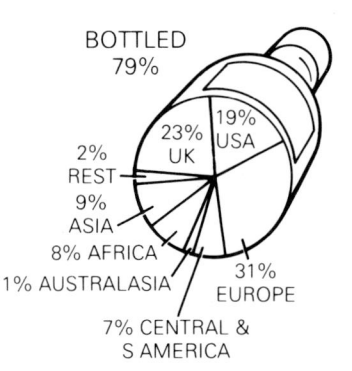

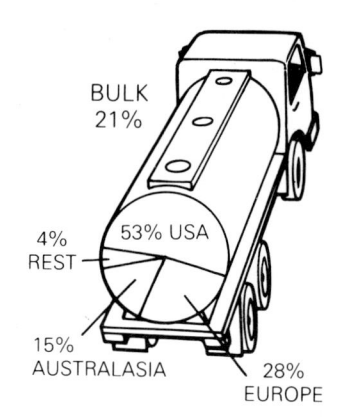

TOTAL
26% USA
18% UK
2% REST
8% ASIA
6% AFRICA
4% AUSTRALASIA
6% CENTRAL & S AMERICA
30% EUROPE

BOTTLED 79%
19% USA
23% UK
2% REST
9% ASIA
8% AFRICA
1% AUSTRALASIA
7% CENTRAL & S AMERICA
31% EUROPE

BULK 21%
53% USA
4% REST
15% AUSTRALASIA
28% EUROPE

Source: The Scotch Whisky Association

quickly the main features of the data. As an example of a pictogram, consider the data on British Council Expenditures on Overseas Aid in 1984/85 displayed in Table 2.6, p. 10. To simplify the data and render it more suitable for representation by means of a pictogram we shall first express the expenditures to the nearest five million pounds, as given in Table 12.1 (derived from Table 2.6). Using the symbol £ to denote an expenditure of five million pounds, a pictogram for the data in Table 12.1 is shown in Figure 12.2. Recording the expenditures to the nearest five million pounds and taking the symbol £ to represent this amount avoids the complication of having to include fractional parts of the symbol £.

Whereas the pictogram in Figure 12.2 gives a good indication of the expenditures in the various areas, it is less informative than the bar diagram shown in Figure 2.5, p. 10 in which a *scaled* axis is used for expenditure.

Table 12.1 *British Council Expenditures on Overseas Aid in 1984/85.*

Area of expenditure	Expenditure*
English Language	40
Science	20
Education	20
Arts	15
Media	5

* To nearest £5m

Figure 12.2 *Pictogram for the Data in Table 12.1*

English Language	£ £ £ £ £ £ £ £
Science	£ £ £ £
Education	£ £ £ £
Arts	£ £ £
Media	£

KEY
£ = £5m

● **EXERCISE 12.1**

1 Look through newspapers (such as The Times, The Daily Telegraph, and The Guardian) for enhanced pictorial representations or pictograms of statistical data. In each case where sufficient information is provided, represent the data in one of the diagrammatic forms described in earlier chapters.

2 Draw a pictogram to illustrate the methods of travelling to school used by the 20 boys (or the 16 girls) as given in Table 2.9, p. 12 (The symbol ⚹ is one possible way of representing one boy (or girl).

12.2 Misleading Diagrams

In Section 2.2, p. 12 we made the point that when a closed figure is used to represent a numerical value, then it is its *area* that should be proportional to the numerical value. To do otherwise can give a misleading impression. Figure 12.3 shows a diagrammatic representation of the amounts to which an investment of £1000 made in 1975 had grown in 1980 and 1985.

Figure 12.3 *Growth of an Investment Worth £1000 in 1975*

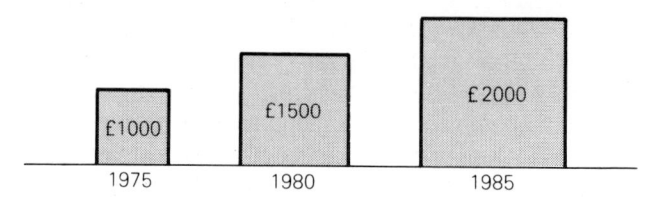

£1000 (1975)   £1500 (1980)   £2000 (1985)

In this diagram the length of the side of the square above 1985 is twice that of the square above 1975 so that the area of the square above 1985 is four times that of the square above 1975. Judging areas when comparing squares, as one would normally do, the diagram gives the impression that the investment has quadrupled from 1975 to 1985. However, the figures indicated within the squares show that the investment has only doubled (not quadrupled) over the period.

Figure 5.8, p. 61 is another example in which a diagram consisting of rectangles can give a false impression if the rectangles do not have areas proportional to the numerical values they represent. Similar false impressions arise when solid (three dimensional) figures are used to represent numerical values.

In earlier chapters we emphasized the importance of correctly scaling any axis used in a diagram. This is occasionally abused, especially in advertisements, in order to exaggerate some feature of the data. As an extreme example, consider the data in Table 12.2 of the annual turnover of a manufacturing company in each of several years.

Table 12.2 *Annual Turnover of a Company*

Year	Turnover (£m)
1979	24
1980	25
1981	26
1982	28
1983	32
1984	27
1985	34

A correct time series graph of this data is shown in Figure 12.4. It is evident from the graph that the turnover of the company has increased steadily but not spectacularly, apart from the dip from 1983 to 1984. The graph is easily distorted to make the growth of the turnover appear to be more impressive than it really is. One possibility is shown in Figure 12.5. There are a few points to note about this graph.
(1) Doubling the scale and suppressing the zero point along the vertical axis has made the lines joining successively plotted points steeper, giving the impression that the growth trend is much greater than it really is. (Suppressing the zero point is in order provided that there is some clear indication in the diagram that this has been done. We did this in earlier diagrams (see, for example, those in Chapter 9 in which we

Figure 12.4 *Time Series Plot of the Data in Table 12.2*

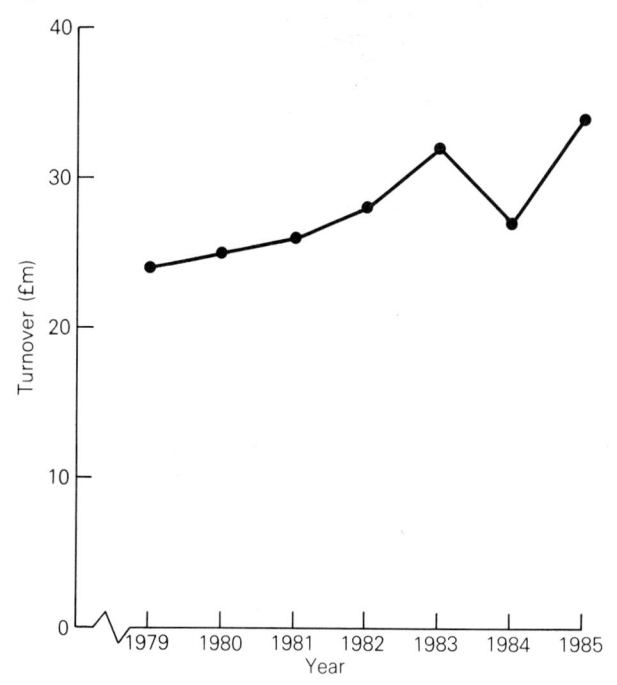

Figure 12.5 *Distorted Graph of the Data in Table 12.2*

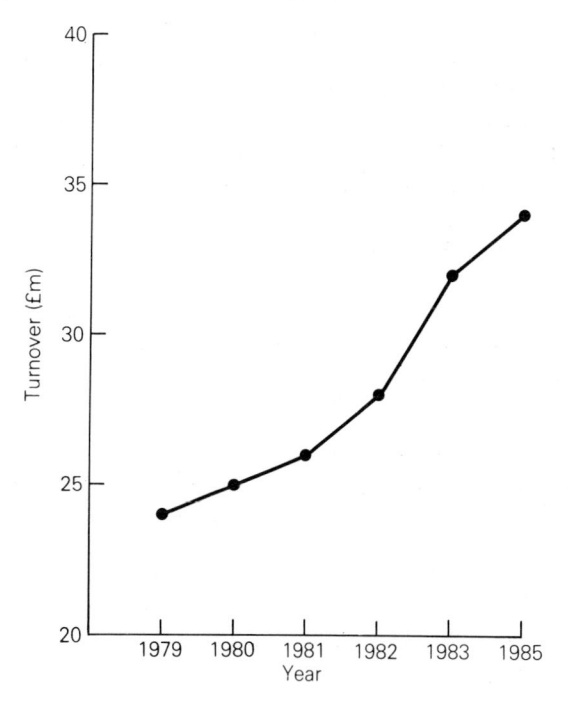

used a zig-zag line to draw attention to the suppression of the zero point).
(2) Omitting the 1984 figure so as not to show that there was a drop in the annual turnover from 1983 to 1984 is a clear case of cheating. Also, along the horizontal axis the two-year period from 1983 to 1985 has the same gap as the one-year periods. The effect of this is to exaggerate the increase from 1983 to 1985. With a properly

[157

scaled horizontal axis the gap from 1983 to 1985 should be twice that from one year to the next. Finally, it would have been more honest to draw attention to the fact that the 1984 figure is not shown, by, for example, using a broken line to join up the 1983 and 1985 figures.

Admittedly we have chosen an extreme example here to demonstrate what may be done. Nevertheless, it does bring out the need to look carefully at any graph. In particular, check that the axes are scaled properly and whether the zero point has been suppressed.

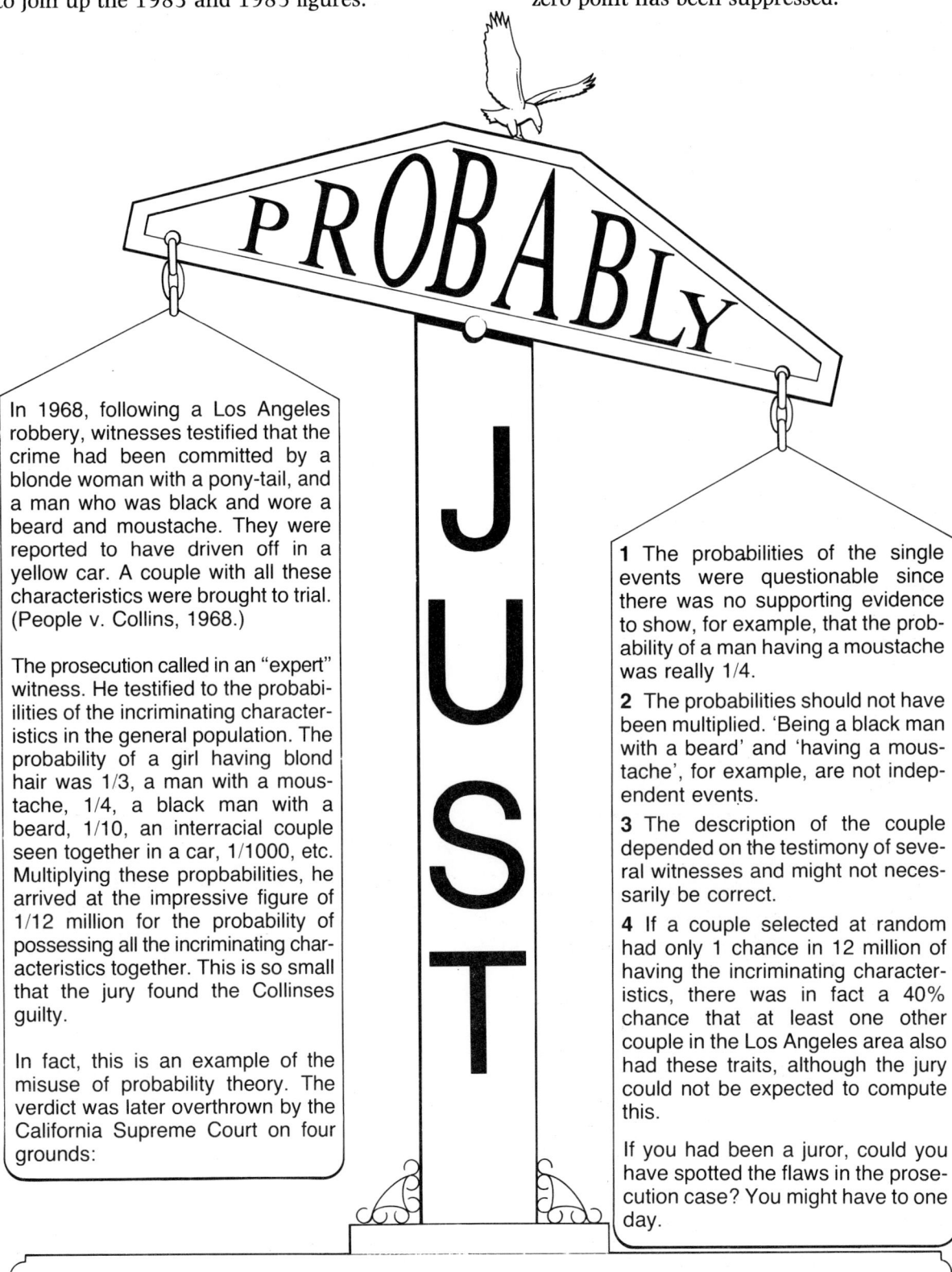

PROBABLY JUST

In 1968, following a Los Angeles robbery, witnesses testified that the crime had been committed by a blonde woman with a pony-tail, and a man who was black and wore a beard and moustache. They were reported to have driven off in a yellow car. A couple with all these characteristics were brought to trial. (People v. Collins, 1968.)

The prosecution called in an "expert" witness. He testified to the probabilities of the incriminating characteristics in the general population. The probability of a girl having blond hair was 1/3, a man with a moustache, 1/4, a black man with a beard, 1/10, an interracial couple seen together in a car, 1/1000, etc. Multiplying these propbabilities, he arrived at the impressive figure of 1/12 million for the probability of possessing all the incriminating characteristics together. This is so small that the jury found the Collinses guilty.

In fact, this is an example of the misuse of probability theory. The verdict was later overthrown by the California Supreme Court on four grounds:

1 The probabilities of the single events were questionable since there was no supporting evidence to show, for example, that the probability of a man having a moustache was really 1/4.

2 The probabilities should not have been multiplied. 'Being a black man with a beard' and 'having a moustache', for example, are not independent events.

3 The description of the couple depended on the testimony of several witnesses and might not necessarily be correct.

4 If a couple selected at random had only 1 chance in 12 million of having the incriminating characteristics, there was in fact a 40% chance that at least one other couple in the Los Angeles area also had these traits, although the jury could not be expected to compute this.

If you had been a juror, could you have spotted the flaws in the prosecution case? You might have to one day.

● **EXERCISE 12.2**

1 A class consists of 20 girls and 20 boys. Figure 12.6 is supposed to be a diagrammatic representation of the numbers of girls and boys in the class who own bicycles. Explain why this diagram is misleading. Draw a diagram to represent the information correctly.

Figure 12.6

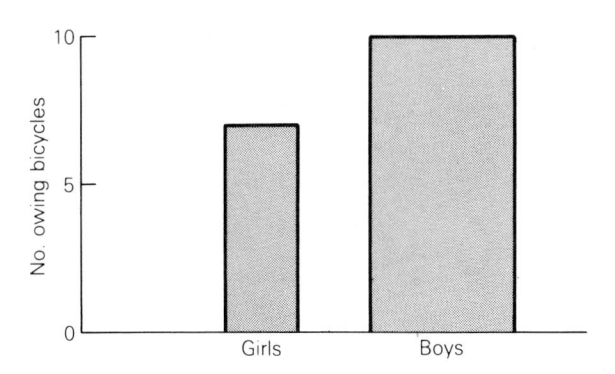

2 Briefly explain why the diagrams in Figures 12.7 and 12.8 may be misleading.

Figure 12.7 *Profits*

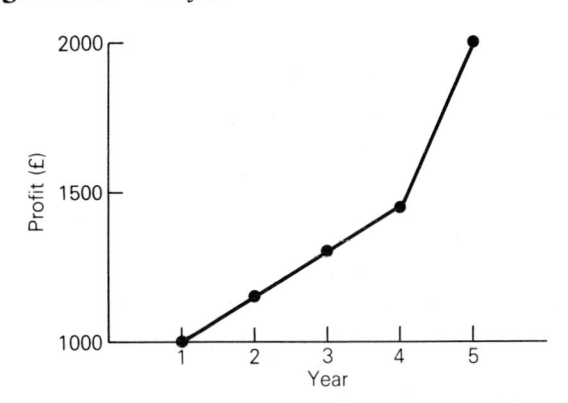

Figure 12.8 *Sales*

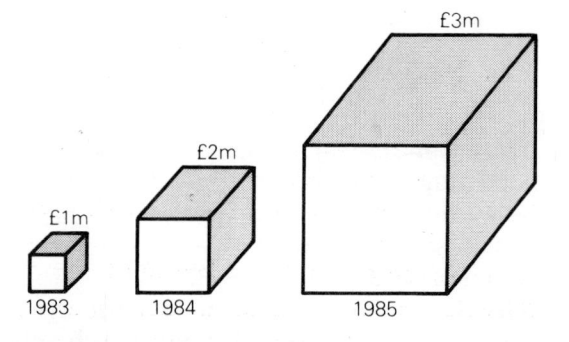

3 The political party in power prior to a general election produced the graph in Figure 12.9 to show the movement of the RPI (retail price index) during their period in office, suggesting that the government had successfully controlled the cost of living. The opposing party produced the graph in Figure 12.10 which suggests that the cost of living had increased substantially during the government's period of office. Comment on the ways in which these graphs have been presented. Explain why both graphs could have been obtained from the same set of data on the retail price index during the four years.

Figure 12.9

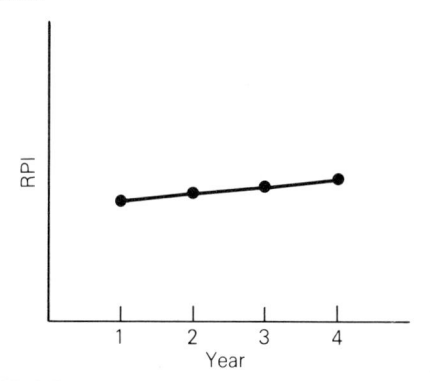

Figure 12.10

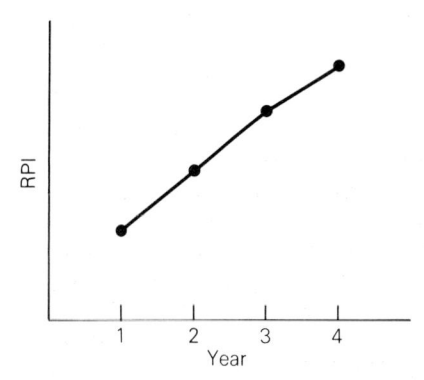

4 Criticize the graph in Figure 12.11 which shows the average amounts of pocket money received by pupils of various ages.

Figure 12.11

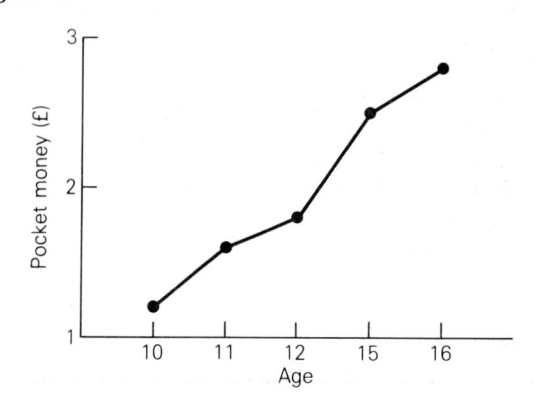

12.3 Ambiguous Statements

Claims made by advertisers and others are sometimes designed to impress but do not stand up to close scrutiny. Some examples are as follows:
(1) '9 out of 10 dogs preferred Lush.'
 This does not say much. Some obvious questions to ask are:
 (a) How many dogs were used?
 (b) Was a fair trial conducted? In particular, what alternatives to Lush were offered to the dogs?
 (c) How hungry were the dogs at the time?
(2) 'Brand A tyres have up to twice the lifetime of the average of all other brands.'
 An analysis of the claim made here shows that little positive is being said about the lifetimes of tyres of Brand A. Firstly, 'up to' means that the actual lifetime of a Brand A tyre could be *less* than the average of the lifetimes of the other brands. Secondly, even if a Brand A tyre does have a lifetime which is twice the average lifetime of tyres of other brands, it does not necessarily follow that a Brand A tyre has a longer lifetime than a tyre of the best of the other brands. (For example, suppose that the lifetime of a Brand A tyre is 32 000 km, and that there are three other brands whose lifetimes are 5000 10 000 and 33 000 km respectively. Then the average lifetime of the other three brands is 16 000 km (exactly one-half of the lifetime of a brand A tyre), but a tyre of the best of the other three brands has a longer lifetime, 33 000 km, than a Brand A tyre).
(3) A driving school advertised in the press as follows: 'Over 500 of our customers have passed their driving tests. Enrol with us to increase your chance of passing'.
 The fact that over 500 customers have passed is not a measure of how successful the instruction has been. For a measure of the driving school's success rate we should take the ratio of the number of successes to the total number of attempts, but the advertisement does not tell us how many attempts (or failures) there have been.

● *EXERCISE 12.3*

Collect advertisers' claims made in newspapers, magazines or on television, and analyse them critically. Note those that appear to be free from any ambiguity and summarize your criticisms of the others.

12.4 Misinterpretations

Misinterpretations of statistical data occur too frequently. The following examples illustrate some types of misinterpretations.
(1) A university's records showed that 16% of all female students and 24% of all male students had contacted the university's medical officer complaining of feeling depressed. It follows that, on average $\frac{1}{2}(16+24)\% = 20\%$ of all the students contacted the medical officer complaining of feeling depressed.

Accepting that the university's records are correct, the error here is due to adding (and averaging) two percentages which may have different bases. To illustrate this, consider three cases in which there are a total of 5000 students consisting of (a) 1000 females and 4000 males, (b) 4000 females and 1000 males, (c) 2500 females and 2500 males.

For (a) the total number of students who had contacted the medical officer complaining of feeling depressed is

(16% of 1000) + (24% of 4000)
$$= 160 + 960 = 1120,$$

so that the overall percentage is:

$$\frac{1120}{5000} \times 100 = 22.4\%.$$

In case (b) the total number is:

(16% of 4000) + (24% of 1000)
$$= 640 + 240 = 880,$$

and the overall percentage is:

$$\frac{880}{5000} \times 100 = 17.6\%.$$

Finally, in case (c), the total number is:

(16% of 2500) + (24% of 2500) = (40% of
$$2500) = 1000,$$

and the overall percentage is:

$$\frac{1000}{5000} \times 100 = 20\%.$$

It is only in case (c) that the averaging of the two percentages gives the correct answer. This is because the numbers of females and males are equal, so that the two percentages have a common base.
(2) Of the 130 motoring accidents that occurred in a certain locality in a period of one month only 15

of them involved women drivers. This shows that women are safer drivers than men.

The conclusion drawn is not justified since it disregards the fact that there are far more men drivers than women drivers. For a realistic comparison of the safety records of women and men we should be looking at the *proportions* of women drivers and of men drivers who were involved in accidents.

(3) Medical records have shown that a certain surgical operation has a success rate of 1 in 5. A patient due to undergo this operation felt much happier when told that the previous four operations had been unsuccessful.

A success rate of 1 in 5 means that the probability of a successful outcome is $\frac{1}{5}$. That is, $\frac{1}{5}$ of a very large number of operations will be successful. The chance that a particular operation will be successful remains constant at $\frac{1}{5}$ whatever the outcomes of preceding operations. Thus the patient in question is being unjustifiably optimistic.

● EXERCISE 12.4

Criticize each of the following.

1 Last year it was found that 3468 pedestrians under the influence of alcohol were injured in road accidents, and that 10 228 car drivers under the influence of alcohol were injured in road accidents. These figures show that a car driver under the influence of alcohol is almost three times more likely to be injured in a road accident than a pedestrian under the influence of alcohol.

2 A person has decided to take a holiday abroad next July and has shortlisted two resorts. The person's preference is to have a holiday where the temperature is 30°C or higher. Records for the two resorts quote the average July temperatures to be 27°C and 32°C. The person chooses the resort whose average July temperature is 32°C.

3 In a certain county 14% of the divorce cases involved men who had married when they were aged between 20 and 30, while only 11% of the cases involved men who had married when they were over 30 years old. It follows that the older a man is when he marries the more likely it is that his marriage will be successful.

4 It has been established beyond any reasonable doubt that a child's reading ability is highly correlated with the length of the child's foot. It follows that in order to improve reading ability attempts should be made to develop treatments that will make feet longer.

5 A pupil carried out a project aimed at finding the average number of children of school age among the families in the school's catchment area. Every pupil in the school was asked to state the number of children of school age in his/her family. The pupil then calculated the mean number per family to be 3.8. Why is this bound to be an overestimate of the true mean number of school age children per family?

6 Having already thrown a fair die twelve times without obtaining a 6, my chance of a 6 in the next throw must be greater than $\frac{1}{6}$.

7 Since most car accidents occur during daytime it is safer to drive at night.

8 The price of a commodity fell 10% from 1983 to 1984 and rose 10% from 1984 to 1985, thus restoring its price to that in 1983.

9 One week apples were sold at 1 kg for 45 pence. The following week the same apples were sold at 3 kg for 45 pence. The average price over the two-week period was 2 kg for 45 pence.

10 When a fair coin is tossed the probability of a head is $\frac{1}{2}$. When a fair die is tossed the probability of an even score is $\frac{1}{2}$. It follows that if a fair coin and a fair die are tossed together the probability of getting either a head or an even score is $\frac{1}{2}+\frac{1}{2}=1$. (Use a tree diagram to find the probabilities of (a) a head or an even score or both, (b) a head or an even score but not both.)

A final word

You will undoubtedly have heard the claim that 'There are lies, damned lies, and statistics'. One purpose of this chapter (and indeed the whole of this book) has been to show you that statistics is a perfectly respectable and relevant subject if the methodology is used properly. The trouble is that some people with very little knowledge of Statistics attempt to interpret data, and others are downright dishonest by manipulating data to suit their own needs.

ANSWERS TO EXERCISES

1 Introducing Data and Variables

EXERCISE 1.1

1. b, c, d, f, h, i, j, l, m, n
2. (i) b, c, f, h, i, l, m, n (ii) d, j
3. a, e, g, k
4. (a) Colour, Type (b) Number of doors (c) Engine capacity, Maximum speed

EXERCISE 1.2

1. (i) b, d, f, h, i, j, l, m (ii) c, n
2. (i) e, g, (ii) a, k

2 Data on a Qualitative Variable

EXERCISE 2.1

1.

Method of travel	Number of pupils
Bus	16
Car	6
Walk	10
Other	4
Total	36

Modal method of travel: Bus.

2.

Season of birth	Number of pupils
Spring	12
Summer	12
Autumn	7
Winter	5
Total	36

3. 35 vowels, 48 consonants, Percentage of vowels = 42.2%

5. (a)

Car colour	No. of cars
Red	2
White	2
Blue	4
Yellow	1
Black	1
Total	10

Modal colour: Blue

(b)

Car type	No. of cars
Saloon	5
Hatchback	3
Estate	2
Total	10

Modal car type: Saloon

EXERCISE 2.2

1. (c) Sector angles: 160° (Bus), 100° (Walk), 60° (Car), 40° (Other)
2. Sector angles: 126° (Bus), 126° (Walk), 54° (Car), 54° (Other)
3. Sector angles: 180° (Saloon), 108° (Hatchback), 72° (Estate)
4. Sector angles: $202\frac{1}{2}°$ (Bus), $67\frac{1}{2}°$ (Car), $67\frac{1}{2}°$ (Walk), $22\frac{1}{2}°$ (Other)
5. Sector angles: 113° (Woman's Own), 93° (Woman), 80° (Woman's Weekly), 74° (Family Circle)

EXERCISE 2.3

1. (b) Sector angles: 166° (U.K.), 115° (North America), 54° (Japan), 25° (Other)
2. (b) Sector angles: 192° (England), 116° (Scotland), 31° (Wales), 21° (Northern Ireland)
3. (b) Sector angles: 299° (Coal), 46° (Coke), 15° (Other)

EXERCISE 2.4

1. Radius of boys' circle to be $1\frac{1}{2}$ times that of the girls' circle.
   Sector angles for boys: 196° (Satisfactory), 104° (Good), 60° (Poor)
   Sector angles for girls: 180° (Satisfactory), 108° (Good), 72° (Poor)
2. Radius for Model B about 1.07 times that for Model A.
   Model A sector angles: $175\frac{1}{2}°$ (Depreciation), 124° (Petrol and Oil), $40\frac{1}{2}°$ (Tax), 20° (Servicing)
   Model B sector angles: 158° (Depreciation), 129° (Petrol and Oil), 44° (Tax), 30° (Servicing)

Review Exercises on Chapter 2

1. (a) 30 (b) Dancing
   (c) Sector angles: 108° (TV), 84° (Reading), 48° (Sport), 120° (Dancing)
2. (c) Sector angles: 108° (Heart), 108° (Club), 72° (Diamond), 72° (Spade)
3. Sector angles: 200° (Computer games), 80° (Table-top games), 50° (Video consoles), 30° (Video games)
4. Sector angles: *either* 168° (Semi-), 120° (Detached), 72° (Link). *or* 108° (Blue), 84° (Brown), 84° (Green), 48° (Red), 36° (Yellow)
6. (b) Sector angles: 212° (Education), 46° (Highways), 38° (Police), 36° (Social Services), 29° (Other)
7. Sector angles: 219° (Conservative), 116° (Labour), 12° (Other), 9° (Liberal), 3° (SDP)
8. (b). Sector angles: 116° (Labour), 101° (Conservative), 55° (Alliance), 48° (Other), 39° (Don't know)
   (d) $2\frac{1}{4}$ cm

3 Small Numerical Data Sets

EXERCISE 3.2

1. (a) Mode=6, mean=5 (b) mode=8, mean=$8\frac{2}{3}$
   (c) mode=9, mean=10 (d) mode=9, mean=13
   (e) mode=1.5, mean=2.5 (f) No mode, mean=242
   (g) No mode, mean=97
2. x=106, mode=106
3. Yes 4. No (not an observed value)
5. Yes 6. No (too large because of the extreme value 31) 7. Yes 8. Yes 9. No (too small because of the extreme value 25)
10. 5 11. 8 12. 10 13. 10 14. 2.65 15. 235 16. 116

EXERCISE 3.3

1. range =4, MAD($\bar{x}$)=MAD(m)=$1\frac{1}{9}$
2. range=4, MAD($\bar{x}$)=$1\frac{2}{9}$, MAD(m)=1
3. range=3, MAD($\bar{x}$)=MAD(m)=$\frac{6}{7}$
4. range=22, MAD($\bar{x}$)=$5\frac{2}{7}$, MAD(m)=$3\frac{6}{7}$
5. range=1.8, MAD($\bar{x}$)=MAD(m)=0.6
6. range=60, MAD($\bar{x}$)=18.4, MAD(m)=17
7. range=100, MAD($\bar{x}$)=28.8, MAD(m)=23.4
8. 1.33 9. 1.37 10. 1.07
11. 7.43 12. 0.68 13. 21.59
14. 36.84
22. $\frac{1}{2}(6-4)=1$ 23. $\frac{1}{2}(10-8)=1$
24. $\frac{1}{2}(11-9)=1$
25. $\frac{1}{2}(12-9)=1\frac{1}{2}$ 26. $\frac{1}{2}(3.15-1.8)=0.675$
27. $\frac{1}{2}(250-225)=12.5$
28. $\frac{1}{2}(118-101)=8.5$

EXERCISE 3.4

7. (a) mean=5260, SD=135.65 (2 d.p.)
   (b) mean=2.375, SD=0.53 (2 d.p.)
9. (a) (i) $\frac{1}{9}, -\frac{7}{9}, -\frac{2}{9}, \frac{15}{9}, \frac{6}{9}, \frac{5}{9}, -\frac{14}{9}, -\frac{3}{9}, -\frac{12}{9}, \frac{11}{9}$
   (ii) 20.2, 18.6, 19.6, 23.0, 21.2, 21.0, 17.2, 19.4, 17.6, 22.2.
   (b) $y=2(x+2)$ 10. Physics
11. (a) First B, Second E (b) First C, Second B

Review Exercises on Chapter 3

1. m=51, $\bar{x}$=50, MAD($\bar{x}$)=12.75, SD=15.07 (2 d.p.)

2 (a) mode=£6000, m=£6000, $\bar{x}$=£9600 (b) £6000 (mode or median)

3 (a) mode=41, m=40.5, $\bar{x}$=40.2, MAD($\bar{x}$)=2.0, SIQR=$\frac{1}{2}$(41−38)=1.5 (b) either $\bar{x}$=40.2, MAD($\bar{x}$)=2.0, or m=40.5, SIQR=1.5.

4 (b) mode=4, m=4, SD=2.91 (2 d.p.), MAD(m)=$2\frac{3}{11}$

5 (a) mode=11, m=11, $\bar{x}$=13 (b) LQ=10, UQ=13, SIQR=1.5 (c) median

6 $\bar{x}$=0.8, SD=2.85 (2 d.p.) **7** 4.08 (2 d.p.) **8** (a) $\bar{x}$=75, SD=30 (b) $\bar{x}$=51, SD=20

9 (a) $\bar{x}$=23.25, VAR=56.1875 (b) (i) 15.23 (ii) $\bar{x}$=27, SD=7

10 (a) (i) 17.38 (ii) 17.37 (iii) 4 min (iv) 4.39 min (2 d.p.) (v) $\frac{1}{2}$(41−33)=4 min (b) either m=17.38 and SIQR=4, or $\bar{x}$=17.37 and MAD($\bar{x}$)=4, or $\bar{x}$=17.37 and SD=4.39

11 (a)(i) 4.07 (ii) 0.00114 (b) y=100(x−0.07)=100x−7

12 x=13, a=3 or 9

13 A, B, C (in rank order)

4 Discrete Frequency Distributions

EXERCISE 4.1

1

Shoe size	5	$5\frac{1}{2}$	6	$6\frac{1}{2}$	7	$7\frac{1}{2}$	8
No. of boys	3	3	2	3	3	4	2

2

Shoe size	4	$4\frac{1}{2}$	5	$5\frac{1}{2}$	6	$6\frac{1}{2}$
No. of girls	2	3	5	3	1	2

3

No. of boys	0	1	2	3	4
No. of families	5	12	13	4	2

4

No. of children	1	2	3	4	5	6
No. of families	4	11	11	7	2	1

(Assume that there are no brother and sisters in the class)

EXERCISE 4.2

1 $7\frac{1}{2}$ **2** 5 **3** 2 **4** No unique mode **5** $6\frac{1}{2}$ **6** $5\frac{1}{8}$ **7** $1\frac{11}{18}$ **8** $2\frac{31}{36}$ **9** $6\frac{1}{2}$ **10** 5 **11** 2 **12** 3 **13** 0.85 **14** $\frac{19}{32}$ **15** $\frac{277}{324}$ **16** $\frac{203}{216}$ **17** 0.99 (2 d.p.) **18** 0.74 (2 d.p.) **19** 1.03 (2 d.p.) **20** 1.18 (2 d.p.) **21** 0.85 **22** $\frac{9}{16}$ **23** $\frac{5}{6}$ **24** $\frac{11}{12}$ **25** $\frac{1}{2}(7\frac{1}{2}-5\frac{1}{2})$=1 **26** $\frac{1}{2}(5\frac{1}{2}-4\frac{1}{2})$=$\frac{1}{2}$ **27** $\frac{1}{2}$(2−1)=$\frac{1}{2}$ **28** $\frac{1}{2}$(4−2)=1 **29** (a) m=2 (b) SIQR=$\frac{1}{2}$(4−2)=1 (c) Last inverval open-ended

EXERCISE 4.3

1 $m=6\frac{1}{2}$, LQ=$5\frac{1}{2}$ UQ=$7\frac{1}{2}$ **2** m=5, LQ=$4\frac{1}{2}$, UQ=$5\frac{1}{2}$

3 m=2, LQ=1, UQ=2 **4** m=3, LQ=2 UQ=4

5 m=2, LQ=2, UQ=4

EXERCISE 4.4

1 (a) m=17 (b) $18\frac{1}{2}-15=3\frac{1}{2}$

2 D_3=4, D_4=D_5=5, D_6=6 (b) D_6-D_4=1

3 P_{40}=2, P_{60}=3

EXERCISE 4.5

1 Negatively skew **2** Almost symmetrical **3** Almost symmetrical **4** Positively skew

5 (b) Positively skew, (c) mode=11, m=11, $\bar{x}$=14.5 (mean to the right of the median) (d) m=11, SIQR=$\frac{1}{2}(13\frac{1}{2}-10\frac{1}{2})$=$1\frac{1}{2}$

6 (b) Almost symmetrical (c) mode=46, m=45, $\bar{x}$=45.42 (2 d.p.); They do not differ very much.

Review Exercises on Chapter 4

1 (a) 24 (b) 2 (c) 2 (d) $1\frac{5}{6}$ (e) $\frac{5}{6}$ (f) median and MAD(m)

2 (a)

No. of goals	0	1	2	3	4
No. of matches	4	11	4	3	2

(b) $\bar{x}$=1.5, SD=1.15 (2 d.p.)

3 (a)

Score	65	66	67	68	69	70
Frequency	2	6	9	12	7	4

(c) m=68, SIQR=$\frac{1}{2}$(69−67)=1
(d) MAD(m)=1.05

4 (a) Both are $1\frac{1}{6}$, (b) $\bar{x}$=5.5, SD=1.5, **5** m=2, SIQR=$\frac{1}{2}(2\frac{1}{2}-1)$=$\frac{3}{4}$

6 (a) 1 (b) $\bar{x}$=1.9, MAD($\bar{x}$)=1.378

7 (a)

No. of errors	0	1	2	3	4+
No of pages	18	12	11	6	3

(b) 1

(c) Open-ended interval (d) SIQR=$\frac{1}{2}(2-0)$=1

8 (a) $\bar{x}$=10.23, SD=1.44 (b) (i) $\bar{x}$=15.23, SD=1.44 (ii) $\bar{x}$=20.46, SD=2.88 (iii) $\bar{x}$=30.46, SD=2.88 (b) m=10, $P_{60}-P_{40}$=11−10=1

9 (a)

No. of times down	0	1	2	3	4	6	7
No. of machines	8	22	4	2	2	1	1

(b) mode=1, median=1, mean=1.425 (c) mean (non-integer)

10 (a) Negatively skew (b) greater than 61
11 (a) $\bar{x}$=1.4, MAD($\bar{x}$)=0.61, SD=0.7 (b) MAD($\bar{x}$)=0.55, SD=0.5.
12 (a) mode=200, m=200, $\bar{x}$=290. SD=181.38 (2 d.p.); (b) mean and standard deviation (c) mode=204, m=204, $\bar{x}$=295.8, SD=185.01 (2 d.p.)
13 (a) $\bar{x}$=15.2625, SD=2.1780 (4 d.p.) (b) English (standardized score in English=0.20, in Mathematics=0.15)

5 Grouped Frequency Distributions

EXERCISE 5.1

1 174.5 cm to 175.5 cm **2** 12.5°C to 13.5°C **3** 58.55 kg to 58.65 kg
4 3 min 58.555 s to 3 min 58.565 s **5** 15.00 to 16.00

EXERCISE 5.2

1

Actual height (cm)	Frequency
158.5–	1
161.5–	3
164.5–	4
167.5–	4
170.5–	6
173.5–	2
176.5–	0

2

Consumption (km per litre)	Frequency
9.75–	3
10.25–	5
10.75–	4
11.25–	7
11.75–	11
12.25–	7
12.75–	3
13.25–	0

3 Positively skew
4 Positively skew
5 Approximately symmetrical

EXERCISE 5.3

1 170.5–173.5
2 11.75–12.25
3 10–11 **4** 22–28 **5** 31–40 **6** 13–14

EXERCISE 5.4

(Answers obtained graphically by you may differ slightly from those given here.)
1 (c) m=29.75, LQ=24.3, UQ=37.2, P_{40}=27.4, P_{60}=32.25 (d) 51%
2 (b) In hundreds of hours: m=10.9, LQ=10.1, UQ=12.4 (c) 23%
3 (a) m=48.8. LQ=36.5, UQ=61.2 (b) 27%

EXERCISE 5.5

Refer to the answers given in Exercise 5.4.

EXERCISE 5.6

1 $\bar{x}$=168.55 cm, SD=4.17 cm **2** $\bar{x}$=11.64 km per litre, SD=0.84 km per litre
3 $\bar{x}$=11.33 hundred hours, SD=1.82 hundred hours
4 $\bar{x}$=31.66 years, SD=9.93 years
5 $\bar{x}$=48.78, SD=18.79
6 $\bar{x}$=13.92 thousand pounds, SD=1.53 thousand pounds

7

Time (min)	60–	65–	70–	75–	80–	85–
Frequency	5	11	8	22	4	0

(b) $\bar{x}$=73.40 min, SD=5.80 min

[165

EXERCISE 5.7

(Answers obtained graphically may differ slightly from those given here.)

1 (b) m=£96, SIQR=£3 (c) $\bar{x}$=£96.18, SD=£6.45
2 (b) m=47.1, SIQR=11.3 (c) P_{40}=43.0, P_{60}=51.3 (d) 86%

Review Exercises on Chapter 5

Answers obtained graphically may differ slightly from those given here.)

1 (a) 45.5–46.5

(b)

Duration (s)	Frequency
40.5–	3
100.5–	5
160.5–	8
220.5–	12
280.5–	13
340.5–	9
400.5–	0

(d)

Duration (s)	Cumulative frequency
100.5	3
160.5	8
220.5	16
280.5	28
340.5	41
400.5	50

(e) (i) 265.5 (ii) 65 (iii) 27%

2 (b) 3, 7, 11, 15, 19, 23 (c) $\bar{x}$=14 miles, SD=5 miles
3 (a) 9.5–24.5 s, (b) 54.5–69.5 s, (d) (i) 52.6 s (ii) 11.4 s, (iii) 44
4 (a) m=£124, SIQR=£23 (b) 8
5 (a) 213.05–215.05, 215.05–217.05 etc. (b) (i) m=218 s, SIQR=1.3 s (ii) 5 (iii) 4 (c) 216 s
6 $\bar{x}$=390 kg, SD=158 kg
7 (a) m=20% (b) *either* SIQR=2.5 *or* 10–90 percentile range=10% (c) (i) P_{40}=18.8%, P_{60}=20.8% (ii) 33% (iii) 35%
8 (a) m=10y 11m, SIQR=3y 2m (b) $\bar{x}$=10y 6m, SD=3y 8m
9 (a) Positively skew (b) (i) 25 min (ii) 10.36 min (iii) 8 min (iv) 85

6 Comparing Numerical Data Sets

EXERCISE 6.2

1 Females: m=32.22y, SIQR=17.91y
  Males: m=32.14y, SIQR=15.61y
2 Boys: $\bar{x}$=61.395 kg, SD=3.92 kg (2 d.p.)
  Girls: $\bar{x}$=53.85 kg, SD=6.26 kg (2 d.p.)
3 Males: m=26.68y, SIQR=5.18y
  Females: m=24.03y, SIQR=4.30y

EXERCISE 6.3

1 (a) B (b) A **2** CV_b=6.4%, CV_g=11.6%
3 Greater (CV=31.7% as compared with 25%)
4 1970 (CV_{1970}=32%, CV_{1985}=20.4%)

Review Exercises on Chapter 6

(Answers obtained graphically may differ slightly from those given here.)

1 (b) A: m=123.3, SIQR=20.8, B: m=126.5, SIQR=32.8.
  (c) (i) 41 percentile=115, 98 percentile=285;
  (ii) $P_{60}(B) - P_{60}(A) = 138 - 132 = 6$.
2 $\bar{x}$=£2752, SD=£535.63 (b) CV_{food}=19.46%, $CV_{clothing}$=33.33%.
3 First micrometer (CV_1=0.26%, CV_2=0.47%)
4 Under 5: $\bar{x}$=0.577, SD=0.97, CV=168%
  5 or more: $\bar{x}$=1.471, SD=1.06, CV=72%

7 Populations, Samples and Questionnaires

EXERCISE 7.3

1 57 (11–13), 46 (14–16), 17 (17+)
2 Sex, skill-level, manual or professional

Review Exercises on Chapter 7

9 38, 26, 11, 02, 05, 31, 01, 15, 34, 16, 17, 37, 24, 39.

8 Some Other Summary Measures

EXERCISE 8.1

1 68 **2** (a) 64 (b) 91 **3** £1.28
4 £58 **5** 41.24p

EXERCISE 8.2

1 100, 147, 195, 438, 924, 1408   **2** (a) (i) 120 (ii) 75 (b) £39   **3** (a) £2.50 (b) 132
4 1981: 125, 1982: 200   **5** (a) 150 (b) $66\frac{2}{3}$
6 96, 93.

EXERCISE 8.3

1 104.5   **2** 100, 103.9, 107.7
3 118.2   **4** 100, 109, 104

EXERCISE 8.4

1 (a) 122.87 (2 d.p.) (b) 5.92 (2 d.p.)   **2** (a) $a=3.08$, $b=84$ (b) 99.52   **3** (a) 108%, 110%, 112% (b) 110 (c) 10%   **4** (a) 18 cm, 21.6 cm, 25.92 cm (b) 19.72 cm (2 d.p.)
5 (a) 300, 425, 520, (b) 372

EXERCISE 8.5

1 (a) (i) 14 per '000 (ii) $17\frac{1}{3}$ per '000 (b) (i) 13.62 per '000 (2 d.p.) (ii) 22.13 per '000 (2 d.p.)
2 (a) $10\frac{1}{3}$ per '000 (b) 10.30 per '000 (2 d.p.)
3 (a) 1975: 20.18 per '000 (2 d.p.) 1985 7.11 per '000 (2 d.p.) (b) 7.73 per '000
4 (a) 2.84 per '000, 3.32 per '000 (b) the town
5 6.115%
6 (a) A: 60, B: 99.6, greater number in age range 20–39 (b) A:60, B:60
7 (a) A: 60 per hundred sown, B: 70 per hundred sown
 (b) B, standardized rate with respect to A=68.87 per hundred
8 (a) A: 14.5 per '000 B: 11.125 per '000
 (b) A: 14.46 per '000, B: 11.12 per '000

Review Exercises on Chapter 8

1 (a) 73% (b) 86%
2 (a) 1983: £4760, 1984: £4920, 1985: £5410 (b) 100, 103.36 (2 d.p.), 113.66 (2 d.p.)
3 (a) 114.29 (2 d.p.) (b) 104 (c) 91
4 (a) (i) £4.6 (ii) 80.3 (b) (i) 108 (ii) 102
5 (a) 250 (b) 300 (c) £$3\frac{1}{3}$
6 (a) Coffee and cocoa (b) (i) 320, 328, 362 (ii) 100, 103, 113   **7** 138
8 (a) (i) air (ii) sea (b) 109%, 107%, 106%, 102% (c) 1980 to 1981
9 $x=5$ **10** (a) 119.23, 116.67, 107.14 (2 d.p.) (b) 114.2
11 (a) £1000, £1050, £1102.50, £1157.625 (b) £1077.53, £1075.93 (c) GM
12 (a) (i) 100, 105.0, 115.1, 118.9 (ii) 100, 110.6, 117.2, 122.8 (iii) 100, 98.9, 112.7, 114.5 (b) 5.03%, 9.58%, 3.28%, (2 d.p.) (c) 5.4%
13 112   **14** (a) 8.4 (b) 33.6
15 (a) 20.90 per '000 (2 d.p.) (b) 22.35 per '000 (2 d.p.)
16 (a) 1.2 per '000 (b) 1.26 per '000
17 (a) 22.6 per '000, 22.4 per '000 (b) 26.5 per '000, 21 per '000
18 (a) A: 72 per hundred, B: 75 per hundred (b) 73.2 per hundred (c) C

9 Paired Variables

EXERCISE 9.2

(The answers from your fitted lines may differ slightly from those given here.)

1 $(\bar{x}, \bar{y}) = (5.125, 154.5625)$ (a) 148 cm (b) 164 cm
2 $(\bar{x}, \bar{y}) = (154.5625, 53.85)$ (a) (i) 47.4 kg (ii) 57.5 kg (b) (i) 149 cm (ii) 164 cm
3 $(\bar{x}, \bar{y}) = (60, 10.36)$ (a) (i) 10.1 m (ii) 10.5 m (b) (i) 47.8°C (ii) 75.6°C
4 $(\bar{x}, \bar{y}) = (2.5, 6.4375)$ (a) 6.4 kg (b) 6.9 kg
5 $(\bar{x}, \bar{y}) = (15, 50)$ (a) (i) 53.4 (ii) 47.1 (b) (i) 12.9 (ii) 23.3

EXERCISE 9.3

1 $y=0.7x+4.1$ (a) (i) 50 (ii) 66 (b) (i) 51 (ii) 80
2 $y=0.8x-73.4$ (a) (i) 57 kg (ii) 63 kg (b) (i) 164 cm (ii) 173 cm
3 $y=10.9x+98.7$ **4** $y=0.67x-50.3$
5 $y=0.01x+9.8$ **6** $y=0.94x+4.1$
7 $y=57.2-0.5x$

EXERCISE 9.4

1 (a) $y=0.7989x-73.1853$ (b) 59.44 kg
2 (a) $y=0.009x+9.82$ (b) (i) 10.09 m (ii) 10.495 m
3 (a) $y=-0.48x+57.2$ (b) (i) 53.36 (ii) 47.12

EXERCISE 9.5

1 $\frac{31}{42}=0.738$ (3 d.p.), good agreement
2 $r_S=0.2$, poor agreement
3 A, ($S_A=14$, $S_B=16$)   **4** $r_S=0.41$ (2 d.p.)   **5** $r_S=0.85$ (2 d.p.)   **6** $r_S=0.43$ (2 d.p.)
7 (a) $r_S=-0.77$ (2 d.p.) (b) $r_S=-0.82$ (2 d.p.) (c) $r_S=0.76$ (2 d.p.)

EXERCISE 9.6

1. 0.87 (2 d.p.)   2. (a) −0.88 (2 d.p.), (b) −0.88 (2 d.p.), (c) 0.93 (2 d.p.)
3. 0.90 (2 d.p.)   4. (a) −0.997 (3 d.p.) (b) 0.053 (3 d.p.)

Review Exercises on Chapter 9

1. (b) $\bar{x}=55$, $\bar{y}=1.2075$ (d) (i) 1.25 cm^3, (ii) 0.02 cm^3
2. (b) $\bar{x}=22.5$, $\bar{y}=7.1$ (d) (i) 6.4 (ii) 1.3
3. (b) $\bar{x}=99$, $\bar{y}=104.7$ (c) $y=104$, $x=72$ (d) $r_S=-0.84$ (2 d.p.)
4. $r_S=0.43$ (2 d.p.)   5. $r_S=0.7$
6. A, ($S_A=10$, $S_B=14$)
7. (a) $\bar{x}=9500$ tonnes, $\bar{y}=5000$ tonnes (b) 4900 tonnes, 9500 tonnes (c) $r_S=0.97$ (2 d.p.)
9. (a) $\bar{x}=2.6$m, $\bar{y}=81\,000$ (b) 13.7 (approx.) (c) 91 000
10. (a) $\bar{x}=3.6$ years, $\bar{y}=£476\,000$ (b) $y=7.1-0.7x$ (approx.) (c) £3200 (approx.)
11. (b) $y=2.2x+1$ (approx.) (c) 2.8 kg, 0.6 m
12. $y=0.0045x+0.96$, 1.2525 cm^3
13. (a) $y=-0.134x+10.11$, 6.36 (b) 0.999 (3 d.p.)
14. −0.858   15. $y=0.6736x-50.2627$, 50.8 kg
16. 0.963   17. 0.997 (3 d.p.)

10 Time Series

EXERCISE 10.1

1. (b) Pass rates: 47.3%, 47.3%, 47.6%, 48.1%, 48.7%, 49.0% respectively
2. (b) From the end of the first quarter in 1984 to the end of the first quarter in 1985.
3. (c) mean is at (1966, 74.05) (d) 78.8 years

EXERCISE 10.3

(Graphical answers obtained by you may differ slightly from those given here.)

1. £7530m   2. £125 000
3. (b) use three-point moving averages (c) 230
4. Average increase of £17m
5. (a) +250p (b) +237p
6. (a) +£159m (b) +£143m
7. (a) +£4000 (b) −£333
8. (a) +6 (b) +9

Review Exercises on Chapter 10

(Graphical answers obtained by you may differ slightly from those given here.)

1. (b) mean is at (1966, 68.2) (c) 72 years
2. (b) mean is at (1981, 19.4) (c) 21 million
3. (b) mean is at (1981½, 8.95); 6100
4. (b) 230p (c) General increase, in real terms, from 1976 to 1983, but then a slight drop in 1984.
5. Approximately linear downward trend
6. (b) Approximately linear increasing trend (c) Slope $\simeq 1.2$, on average just over one extra day of snow per year during the period
7. (b) (i) 5.8% (ii) 3.9%
8. (b) Use three-point moving averages (c) £206 000
9. (b) Use four-point moving averages, 13
10. (b) Use six-point moving averages 26.5, 21
11. Use four-point moving averages (a) Decrease of about 0.7 (b) Average increase of about 0.4.
12. (b) Use four-point moving averages (c) (i) £82 000, £112 000 (ii) Average increase of about £400

11 Probability

EXERCISE 11.1

1. (a), (b), (c)

EXERCISE 11.2

1. $\frac{1}{5}$ 2. $\frac{1}{4}$ 3. $\frac{1}{4}$ 4. $\frac{1}{5}$

EXERCISE 11.3

1. (a) (i) $\frac{1}{2}$ (ii) $\frac{3}{10}$ (iii) $\frac{1}{5}$ (iv) $\frac{1}{2}$ (b) $\frac{4}{5}$
2. (a) $\frac{1}{4}$ (b) $\frac{1}{4}$ (c) $\frac{1}{2}$
3. (a) $\frac{1}{6}$ (b) $\frac{7}{30}$ (c) $\frac{1}{3}$ (d) $\frac{1}{6}$ (e) $\frac{1}{10}$ (f) $\frac{2}{5}$ (g) $\frac{3}{5}$
4. (a) (i) $\frac{1}{8}$ (ii) $\frac{11}{40}$ (iii) $\frac{3}{8}$ (iv) $\frac{3}{10}$ (v) 0 (b) $\frac{11}{20}$ (c) $\frac{2}{5}$
5. (a) (i) $\frac{2}{5}$ (ii) $\frac{7}{10}$ (b) (ii) $\frac{1}{10}$ (iii) $\frac{1}{3}$

EXERCISE 11.4

1. $\frac{1}{2}$ 2. (a) $\frac{1}{2}$ (b) $\frac{2}{3}$ (c) $\frac{2}{3}$ 3. (b) (i) $\frac{1}{3}$ (ii) $\frac{1}{3}$ (iii) $\frac{1}{3}$ (c) $\frac{2}{5}$
4. (a) (i) $\frac{1}{12}$ (ii) $\frac{1}{3}$ (iii) $\frac{2}{3}$ (b) $\frac{2}{3}$
5. (a) $\frac{1}{6}$ (b) $\frac{1}{3}$ (c) $\frac{1}{3}$ (d) $\frac{2}{3}$

EXERCISE 11.5

1. (a) $\frac{74}{100}$ (b) $\frac{54}{100}$ 2. (a) $\frac{2}{3}$ (b) $\frac{2}{3}$ 3. $\frac{5}{9}$
4. (a) $\frac{1}{2}$ (b) $\frac{1}{3}$ 5. (i) $\frac{1}{3}$ (ii) $\frac{2}{3}$

EXERCISE 11.6

1 (a) $\frac{1}{8}$ (b) $\frac{1}{2}$ (c) $\frac{1}{2}$ (d) $\frac{1}{2}$ (e) $\frac{1}{4}$ **2** $\frac{2}{3}$ **3** (a) $\frac{15}{28}$ (b) $\frac{27}{64}$
4 (a) 0.729 (b) 0.243 (c) 0.27 **5** (a) 0.63 (b) 0.03

EXERCISE 11.7

1 (a) $\frac{1}{210}$ (b) $\frac{1}{14}$ (c) $\frac{8}{105}$ (a) $\frac{12}{95}$ (b) $\frac{2}{19}$
3 (a) $\frac{1}{4}$ (b) $\frac{5}{36}$ (c) $\frac{5}{18}$ **4** (a) $\frac{1}{8}$ (b) $\frac{1}{54}$ (c) $\frac{25}{72}$
5 (a) $\frac{7}{72}$ (b) $\frac{1}{9}$ (c) $\frac{1}{8}$ **6** (a) $\frac{1}{4}$ (b) $\frac{3}{16}$

EXERCISE 11.8

1 (a) (i) 0.544 (ii) 0.261 (iii) 0.195 (b) (i) 1024 (ii) 492 (iii) 368
2 (b) 12 **3** (a) (i) 0.12 (ii) 0.2 (ii) 0.08 (b) (i) $\frac{1}{9}$ (ii) $\frac{1}{6}$ (iii) $\frac{1}{12}$
4 (a) (i) $\frac{28}{45}$ (ii) $\frac{29}{45}$ (b) (i) 24.9 (ii) 25.8

Review Exercises on Chapter 11

1 No. (a probability cannot be negative)
2 $\frac{3}{4} + \frac{1}{2} \neq 1$
4 $\frac{12}{150}$ **5** $\frac{1}{5}$ **6** (a) $\frac{1}{4}$ (b) $\frac{1}{2}$
7 0.869 **8** $\frac{1}{4}$
9 (a) 0.1; (b) 5 **10** (a) (i) $\frac{4}{15}$ (ii) $\frac{5}{9}$ (iii) $\frac{1}{15}$ (iv) $\frac{32}{45}$ (b) $\frac{2}{5}$ (c) 30
11 (a) $\frac{1}{12}$ (b) $\frac{23}{60}$ **12** (a) $\frac{3}{5}$ (b) $\frac{14}{25}$ **13** (a) (i) $\frac{1}{4}$ (ii) $\frac{3}{4}$; (b) $37\frac{1}{2}$
14 (a) (i) $\frac{2}{3}$ (ii) $\frac{1}{3}$ (b) (i) $\frac{14}{33}$ (ii) $\frac{17}{33}$ (iii) $\frac{16}{33}$ (c) 4
15 (a) $\frac{1}{10}$ (b) $\frac{1}{2}$ (c) $\frac{1}{2}$ (d) $\frac{1}{30}$
16 (a) $\frac{1}{10}$ (b) $\frac{1}{2}$ (c) $\frac{1}{2}$ (d) $\frac{3}{100}$ **17** (a) $\frac{5}{18}$ (b) $\frac{1}{4}$
18 (a) $\frac{5}{6}$ (b) $\frac{1}{12}$ **19** (a) $\frac{1}{10}$ (b) $\frac{3}{10}$ (c) $\frac{3}{5}$
20 (a) 0.2 (b) (i) 44.8 (ii) 6.4 (iii) 12.8

12 Pictograms, Misleading Diagrams, Ambiguous Statements and Misinterpretations

EXERCISE 12.2

1 The areas of the rectangles are not proportional to the frequencies.
2 (a) The suppressed zero exaggerates the profit growth. (b) The volumes of the cubes are not proportional to the sales.
3 No vertical scale is shown.
4 The suppressed zero on the vertical axis exaggerates the increase with age; the horizontal axis is incorrectly scaled.

INDEX

addition rule for probabilities   145
age-specific death rate   94
ambiguous statements   160
arithmetic mean   18, 34, 58
average seasonal effect   136

bar (block) diagram   8
bias   76
bimodal distribution   52
box-and-whisker plot   25

causal association   104
census   75
chance   140
circular diagram   9, 13
class intervals   50
   equal   50
   open-ended   63
   unequal   60
cluster sampling   79
coefficient of variation   72
comparative pie charts   13
comparing examination marks   29
component bar diagram   8
continuous variable   4
convenience sampling   82
correlation   115
crude rates   93
   death   94
   other   96
cumulative frequency   39
   diagram   39, 42, 54
   distribution   39
cumulative relative frequency   40
   diagram   42, 56
cyclical variation   129

diagrams
   bar (block)   8
   box-and-whisker   25
   circular (pie)   9, 13
   comparative   67

component bar   8
cumulative frequency   39, 42, 54
dot   17
dual bar   11
dual percentage bar   12
enhanced   155
frequency bar   33
misleading   156
pie (circular)   9, 13
scatter   103
stem and leaf   17
vertical line   33
discrete frequency distribution   33
   variable   3
distribution shapes   43, 51
dot diagram   17
dual bar diagram   11
dual percentage bar diagram   12

elementary events   141
enhanced diagrams   155
equally likely outcomes   141
events
   elementary   141
   impossible   140
   mutually exclusive   145
   non-elementary   141
   sure   140
expected frequency   152
extrapolation   111

forecasting   127
frequency   7
   bar diagram   33
   density   61
   polygon   51
frequency distribution
   discrete variable   33
   grouped   49
   qualitative variable   7

geometric mean   91
   weighted   92

gradient of a line 112
grouped frequency distribution 49, 60, 63

histogram 50, 61, 68

ideograph 155
impossible event 140
index numbers 88
   simple 88
   weighted (composite) 89, 92
intercept of a line 111
interpercentile range 42
interquartile range 24
interval scale of measurement 5
irregular variation 129

J-shaped distribution 43
judgemental sampling 82

leading question 83
least squares regression line 114
levels of measurement 5
line chart (graph) 126
line fitting by eye 108, 127
line of 'best' fit 114
linear interpolation 56
long-term trend 129
lower quartile 24

mean
   arithmetic 18, 34, 58
   geometric 91
   weighted 86, 92
mean absolute deviation 20
   from the mean 20, 21, 36
   from the median 20, 22, 37
measurement accuracy 48
measures of
   central tendency 18
   dispersion (spread) 20, 36
   location (average) 18, 34
median 18, 19, 35
misinterpretations 160
misleading diagrams 156
modal class 53
modal value 7
mode 18, 34
moving averages 131
multiplication rule for probabilities 146
mutually exclusive events 145

negatively skew distribution 43, 51
nominal scale of measurement 5
nominal scale of measurement 5
non-elementary events 141
non-frequency data 10

non-random sampling 81
normal distribution 51

ogive 54
open-ended class intervals 63
opinion polls 84
ordinal scale of measurement 5

percentiles
   discrete distribution 41
   grouped distribution 56
   linearly interpolated 56
pictogram 155
pie chart 9, 13
population 75
positive correlation 115
positively skew distribution 43, 51
prediction 127, 134
pre-test 82
probability 140
   relative frequency interpretation 151
   tree diagram 146
product moment correlation coefficient 120
pseudo-random digits 76
purposive sampling 82

qualitative variable 4
quantitative variable 2
quartiles 24
   graphically 39, 54
   linearly interpolated 57
questionnaire 82
   example 83
quota sampling 82

random digits 76
   table 77
random sample 76
random sampling procedure 76
random variation in time series data 129
range 20
rank scale of measurement 5
ratio scale of measurement 5
relative frequency 12
   and probability 151
   histogram 68
remainder method 78

sample 75
sample frame 80
sample space 141
sample survey 75
sampling error 84
scatter diagram 103
seasonal effect 136
seasonal (cyclical) variation 129

secular trend   129
selection bias   76
semi-interquartile range   24, 38, 55
skew distribution   43, 51
slope of a line   112
Spearman's rank correlation coefficient   116
standard deviation   22, 23, 36, 58
standardized rates
   death   94
   other   96
standardized scores   28
stem and leaf diagram   17
step polygon   39, 42
strata   79
stratified random sampling   79
study population   80
summary measures   18, 34, 62
sure event   140
symmetric distribution   43
systematic sampling   80

tally chart   6
target population   75
time series data   126
transformations   26, 37, 59
   to given mean and standard deviation   28

tree diagram   143
trend line   127

uncorrelated variable values   115
unequal class intervals   60
unimodal distribution   43
upper quartile   24

variables   2
   continuous   4
   discrete   3
   qualitative   4
   quantitative   2
variance   22
variation
   coefficient of   72
   cyclical   129
   irregular   129
   random   129
   seasonal   129
vertical line diagram   33

weighted arithmetic mean   86
weighted geometric mean   92
weighted index numbers   89, 92
working origin and scale   25, 26